The High Cost of Cheap Coal: The Environmental Price of South Africa's Mining Industry

Shalini

Copyright © 2024 by Shalini

All rights reserved. No part of this book may be reproduced in any man-ner whatsoever without written permission except in the case brief quotations embodied in critical articles and reviews.
First Printing, 2024

Table of Contents

Chapter 1: Introduction 1
    1.1 Background 1
    1.2 Thesis structure 4

Chapter 2: Literature Review 6
    2.1 Acid Rock Drainage 6
        2.1.1 Generation of ARD 6
        2.1.2 Environmental impact 7
        2.1.3 ARD management strategies 9
        2.1.4 Sources of ARD 10
        2.1.5 Active treatment technologies 11
        2.1.6 Passive treatment technologies 14
        2.1.7 Summary of treatment technologies 15
    2.2 Biological sulphur cycle – treatment options 17
    2.3 Biological sulphate reduction 18
        2.3.1 Sulphate reducing bacteria 18
        2.3.2 Bioreactor configurations 22

2.3.3 Effect of operational parameters 23

2.3.4 Biological sulphate reduction treatment technologies 30

2.4 Challenges facing biological sulphate reduction treatment 30

2.4.1 Selection of a suitable electron donor 30

2.4.2 Biological sulphate reduction kinetics 34

2.4.3 Sulphide management 35

2.5 Biological sulphide oxidation 37

2.5.1 Sulphur oxidising bacteria 38

2.5.2 Generation of biologically produced sulphur 42

2.5.3 Biological sulphide oxidation treatment technologies 44

2.6 Floating sulphur biofilm in wastewater treatment 46

2.7 Application of biological technologies to treat ARD in South Africa 49

2.7.1 Paques THIOPAQ™ process 49

2.7.2 Rhodes BioSURE® 49

2.7.3 Integrated Managed Passive (IMPI) process 50

2.8 Current state of research and development 52

2.9 Research rationale and motivation 53

2.9.1 Integration of sulphate reduction and sulphide oxidation in a hybrid LFCR 53

2.10 Research scope 55

2.10.1 Research hypotheses 55

2.10.2 Research objectives 56

2.10.3 Research strategy 56

Chapter 3: Materials & Methods 58

3.1 Experimental setup 58

3.1.1 Microbial cultures 58

3.1.2 Linear flow channel reactor configuration 59

3.1.3 Linear flow channel reactor operating conditions 59

3.1.4 Sampling layout 61

3.2 Analytical methods 61

    3.2.1 Chemical reagents 61

    3.2.2 pH and REDOX potential 61

    3.2.3 Hydrogen sulphide analysis 61

    3.2.4 Sulphate and thiosulphate analysis 62

    3.2.5 Sulphur analysis 63

    3.2.6 Volatile fatty acids 63

3.3 Floating sulphur biofilm (FSB) 64

    3.3.1 Biofilm disruption and harvesting 64

    3.3.2 Elemental analysis 64

3.4 Determining mixing patterns by a dye tracer study 64

3.5 Scanning electron microscopy 65

3.6 Microbial community dynamics 66

    3.6.1 Genomic DNA sampling and 16S rRNA amplicon sequencing 68

    3.6.2 Metagenomic OTU picking and taxonomy alignment 69

    3.6.3 Metagenomic statistical analysis 69

3.7 Data handling 70

    3.7.1 Kinetic calculations 70

Chapter 4: Demonstration of the hybrid LFCR 73

4.1 Introduction 73

4.2. Hydrodynamics of the LFCR 74

4.3 Demonstration of the hybrid LFCR reactor 76

    4.3.1 Experimental approach 76

    4.3.2 Results and discussion 77

4.4 Evaluating the effect of hydraulic residence time 85

    4.4.1 Experimental approach 86

    4.4.2 Results and discussion 86

4.5 Effect of biofilm disruption on process performance 97

  4.5.1 Experimental approach 97

  4.5.2 Results and Discussion. 98

 4.6 Conceptual model of the hybrid LFCR process 100

 4.7 Conclusions 102

Chapter 5: Reactor scale-up and geometry 104

 5.1 Introduction 104

 5.2 Reactor design of the 8 L LFCR 105

 5.3 Hydrodynamics 106

  5.3.1 Experimental approach 106

  5.3.2 Results and discussion 107

 5.4 Effect of hydraulic residence time 108

  5.4.1 Experimental approach 108

  5.4.2 Results and discussion 111

 5.5 Conclusion 131

Chapter 6: Effect of electron donor 133

 6.1 Introduction 133

 6.2 Acetate as an alternative carbon source 135

  6.2.1 Experimental approach 135

  6.2.2 Results and discussion 135

 6.3 Effect of yeast extract and hydraulic residence time 138

  6.3.1 Experimental approach 138

  6.3.2 Results and discussion 139

 6.4 Conclusion 163

Chapter 7: Microbial community dynamics 165

 7.1 Introduction 165

 7.2 Experimental approach 166

7.3 Results and Discussion 167

    7.3.1 Microbial α and β diversity analysis.. 167

    7.3.2 Microbial community composition and relative abundance at phylum level 170

    7.3.3 Microbial classification and distribution at OTU level 173

7.4 Conclusion 186

Chapter 8: Effect of temperature.. 188

8.1 Introduction 188

8.2 Effect of temperature on fluid dynamics 189

    8.2.1 Experimental approach 189

    8.2.2 Results and discussion 189

8.3 Effect of temperature on process performance 192

    8.3.1 Experimental approach 192

    8.3.2 Results and discussion 193

8.4 Dual reactor operation in series 227

    8.4.1 Experimental approach 227

    8.4.2 Results and discussion 229

8.5 Conclusions 236

Chapter 9: Effect of feed sulphate concentration... 238

9.1 Introduction 238

9.2 Effect of sulphate concentration loading 239

    9.2.1 Experimental approach 239

    9.2.2 Results and Discussion 239

9.3 Dual reactor operation in series 262

    9.3.1 Experimental approach 262

    9.3.2 Results and discussion 264

9.4 Conclusion 279

Chapter 10: Conclusions & Recommendations... 281

10.1 Introduction 281

10.2 Final Conclusions... 282

10.3 Recommendations for future work 287

    10.3.1 Enhancement of biological sulphate reduction component 287

    10.3.2 Enhancement of partial sulphide oxidation component: 288

    10.3.3 Development and application of the hybrid LFCR 288

References **290**

Chapter 1

Introduction

1.1 Background

In South Africa, the legacy of historical gold and coal mining and ongoing activities has resulted in the formation of Acid Rock Drainage (ARD) with significant implications on the environment and socio-economics of the country if not addressed (McCarthy, 2011). Effluents emanating from these mining activities are characterised by high levels of acidity, sulphate and heavy metals with low concentrations of organic material (Roman et al., 2008). The quality of the South African freshwater resource is deteriorating mainly as a consequence of salinity coupled with industrial effluent discharge. This is exacerbated by the decline of the local mining industry in recent years, resulting in numerous mine closures and leaving a legacy of financial, social and environmental issues that require urgent management (McCarthy, 2011; Rose, 2013). Early hydrogeological modelling studies predict that the total volume of ARD requiring treatment in South Africa may exceed several hundred ML/day (McCarthy, 2011). In addition, South Africa, known as a water stressed country, faces extended drought periods, placing strain on water supplies nationwide and affecting millions of households. Under these constraints, it becomes increasingly important to preserve the limited freshwater resource available by ensuring efficient strategic water management and treatment.

The potential long-term nature of ARD generation (10s to 100 of years) and worsening predictions of unmet water demand in South Africa has resulted in the need for the development of effective and economically sustainable treatment technologies to address the challenges of ARD (Harrison et al., 2014). Sulphate is a major pollutant occurring in both ARD and industrial effluent waste streams. High concentrations of sulphate increase the salinity of the receiving water bodies, which in turn has major downstream implications on the environment as well as the availability of potable water. Therefore, one of the major aims of ARD treatment in South Africa is the management of sulphate to acceptable levels.

The primary focus on remediation of ARD-contaminated water in South Africa has been on high volume discharges emanating from abandoned underground mine basins, using established active technologies. Largely overlooked, the continuous, low volume ARD discharges from diffuse sources (waste rock dumps, discard heaps and open-pit mining) associated with coal mining, also have a significant impact on the environment due to the large number of sites and wide geographic distribution. Many of these diffuse ARD sources are in rural or peri-urban areas, making traditional active treatment options less applicable and passive or semi-passive approaches more favourable. Passive treatment options, such as wetlands, require less maintenance, fewer skilled operators and have lower operating costs (Sheoran et al., 2010; Skousen et al., 2017). However, these systems (natural and constructed wetlands) are governed by slow, unpredictable kinetics and require extended hydraulic residence times (Zagury et al. 2007; Skousen et al., 2017; Sato et al., 2017), necessitating large land areas. As conventional active systems become more costly and impractical for application at abandoned sites and in remote regions, there has been an increased interest into the development of semi-passive biological systems.

Semi-passive ARD treatment systems are attractive for addressing these low-flow discharges, due to lower capital and operational costs than active systems, as well as improved control and greater predictability than passive systems. Semi-passive systems, based on the action of sulphate reducing bacteria (SRB), can be applied to address sulphate reduction, metal precipitation and wastewater neutralisation simultaneously (Gopi Kiran et al., 2017). Under anaerobic conditions, sulphate reducing bacteria (SRB) reduce sulphate, in the presence of a suitable electron donor, generating sulphide and alkalinity (Zhang & Wang, 2014). Despite extensive research demonstrating the technical feasibility and potential of biological sulphate reduction (BSR) for ARD treatment, relatively few commercial processes have been developed. These technologies have been limited to a few niche applications, due to the relatively slow sulphate reduction kinetics of SRB often constrained by low growth and resultant biomass concentration, high cost of electron donor as well as the management of the sulphide product, which is significantly more toxic than sulphate (Rose 2013; Harrison et al., 2014; van Hille et al., 2015). Each of these challenges must be addressed in order to develop a robust process that can be implemented at an industrial scale.

Due to sulphide toxicity, potential for its re-oxidation, malodour, corrosivity and other hazards, an effective sulphide removal step is essential to ensure satisfactory treatment of sulphate-laden wastewaters (Syed et al., 2006; Harrison et al., 2014). Sulphur oxidising bacteria (SOB) produce elemental sulphur as an intermediate in the oxidation of hydrogen sulphide to sulphate under oxygen limiting conditions (Kleinjan et al., 2005; Cai et al., 2017). Biological sulphide oxidation has been applied in treating sulphide-rich waste streams (Cai et al., 2017), but commercial application has been limited to active treatments, such as the Thiopaq® process (Janssen et al., 2000). Biologically mediated partial oxidation of sulphide to elemental sulphur is restricted to a narrow pH and redox potential range (pH 6 to 8, -20 to -200 mV) (Mooruth, 2013). In addition, the stoichiometric ratio of sulphide to oxygen needs to be maintained at 2:1 to facilitate partial oxidation to elemental sulphur and prevent complete oxidation to thiosulphate and sulphate. This requires precise control of operational conditions, particularly the pH and supply of oxygen (Elkanzi, 2009). Common drawbacks inherent with these active processes are the need for additional operational units and energy requirements, which increase capital and operational costs (Cai et al., 2017). A promising approach to achieve partial sulphide oxidation is through the formation of a floating sulphur biofilm (FSB). This is a passive process that does not require energy input and provides an effective alternative for the removal of sulphide and recovery of elemental sulphur (Molwantwa et al.,2010; Rose, 2013). Floating sulphur biofilms were first observed on waste stabilisation ponds used to manage high sulphide tannery effluents. The biofilm impedes oxygen mass transfer, creating a discrete pH and redox microenvironment that facilitates partial oxidation of sulphide from the bulk liquid, with deposition of the sulphur product in the organic biofilm (Molwantwa et al., 2010; van Hille & Mooruth, 2014).

The application of FSB for sulphide oxidation was first described in the Integrated Managed Passive IMPI process developed by Pulles, Howard and de Lange (Rose, 2013). Degrading packed bed reactors (DPBRs) were used for biological sulphate reduction, followed by linear flow channel reactor (LFCR) units for partial sulphide oxidation via FSB (Coetser et al., 2005). The process was evaluated at demonstration scale but faced several challenges, particularly the sulphide oxidation component, which did not perform optimally (van Hille et al., 2011; van Hille & Mooruth, 2014). A detailed study by Mooruth (2013) led to further optimisation of design and operational parameters of the LFCR. The study demonstrated the feasibility of obtaining high partial oxidation rates in a sulphide-fed linear flow channel reactor (LFCR) through FSB formation.

In a separate study, van Hille et al. (2015) demonstrated the potential application of carbon microfibers as support matrices for biological sulphate reduction within a closed linear flow channel reactor (LFCR). The carbon microfibres facilitated high surface area for biomass retention without significantly reducing the effective reactor volume. The study achieved a high sulphate reduction conversion of 95% at a feed sulphate concentration of 1 g/L. During the study, complete elimination of oxygen was not possible and there was evidence of partial sulphide oxidation and the formation of a FSB. This was similar to that observed in the dedicated sulphide oxidation reactor unit described by Mooruth (2013). This suggested that partial sulphide oxidation could be coupled with sulphate reduction within a single LFCR configuration.

The present study aims to build on these findings through the development of a hybrid LFCR process capable of targeting sulphate removal through the integration of biological sulphate reduction and partial sulphide oxidation, with the recovery of elemental sulphur as a value-end product. During the initial development of any novel treatment, process characterisation for optimal process performance becomes an important criterion. Several studies have focused on improving the overall performance of sulphate reduction and sulphide oxidation by assessing various physicochemical parameters such as concentrations of sulphate, sulphide, pH, choice of electron donor and operational parameters. However, there is a lack of knowledge concerning the influence of these factors on the microbial community that catalyse the biochemical reactions for the desired process. Specifically, the response of the active microbial community to system perturbations and the subsequent effect on the overall process performance and resilience of the system to its fluctuations.

In order to provide a comprehensive approach to the bench-scale demonstration and assessment of the novel hybrid LFCR process, the inter-relatedness of process kinetics, the effects of operating conditions and microbial community dynamics on process performance was investigated.

1.2 Thesis structure

An extensive review of literature is presented in **Chapter 2** with a focus on key aspects of ARD in the context of South Africa, including treatment approaches implemented for its remediation, drawbacks and possible sustainable solutions, as well as biological treatment options such as the application of sulphate reducing and sulphide oxidising bacteria for treating sulphate-rich waste streams. The review highlights the need for the development and validation of economically sustainable passive/semi-passive processes to address the long-term impact of ARD formation and discharge. **Chapter 2** concludes with a detailed description on the research motivation, hypotheses and the set objectives to be addressed in the current research. The analytical procedures, and experimental set-up are detailed in **Chapter 3**.

A broad outline of the experimental chapters are as follows: The initial start-up and lab demonstration of the hybrid LFCR process is presented in **Chapter 4**. This includes the initial start-up phase and characterisation on the effect of hydraulic residence time (HRT) on system performance as well as assessing the effects of managing the floating sulphur biofilm. **Chapters 5 and 6** present the investigations into the effects of reactor geometry (scale-up) and the use of an alternative electron donor (acetate) on process performance, respectively. The experiments were performed in parallel with the effect of hydraulic residence time applied as a key parameter for evaluating the different reactor systems. **Chapter 7** highlights the link between process performance and microbial community dynamics using 16S rRNA amplicon sequencing. This aspect of the investigation was crucial for identification of key microbial community members, implicated in the performance of the process, as well as the effects of hydraulic residence time (HRT) on overall microbial community structure. **Chapters 8 and 9** evaluated the effects of temperature and sulphate loading on process performance, respectively. The focus of these experiments was to determine optimal operating conditions as well as to evaluate process robustness and resilience to recover performance. Lastly, the final conclusions and recommendations from this study are presented in **Chapter 10**.

Chapter 2

Literature Review

2.1 Acid Rock Drainage

Acid Rock Drainage (ARD) is a major environmental issue facing South Africa and is a growing worldwide concern, predominantly associated with the mining of sulphidic minerals. South Africa is renowned for its gold and coal mining industry (McCarthy, 2011; Feris & Kotze, 2014), both having associated sulphidic fractions. The long term legacy of theses mining activities has resulted in ARD pollution that threatens the environment and places increased pressure on the water security of the country. Historically, the potential impact of ARD has been underestimated by mining companies and the government, resulting in the need for the emergency measures currently being implemented to deal with ARD discharge from the underground workings within the Witwatersrand basins (Feris & Kotze, 2014) as well as ongoing environmental degradation associated with diffuse acidic leachates from legacy sulphidic waste rock dumps and tailings facilities.

2.1.1 Generation of ARD

The generation of ARD can be defined as the accelerated oxidation of sulphidic minerals, predominantly iron pyrite (FeS_2) resulting from exposure to oxygenated water as a consequence of their liberation through mining and processing of sulphidic mineral ores and pyritic coal (Johnson & Hallberg, 2005; Kaksonen & Puhakka, 2007). The complete process of pyrite oxidation may be summarised by Reaction 2.1 as follows:

$$4\,FeS_2 + 15\,O_2 + 14\,H_2O \rightarrow 4\,Fe(OH)_3 + 8\,SO_4^{2-} + 16\,H^+ \qquad \text{(Reaction 2.1)}$$

During mining operations, overburden rock is removed in order to gain access to a valuable ore body, creating a network of well-ventilated underground workings as well as exposed waste rock. These exposed mineral rocks may contain sulphide minerals such as FeS_2, the most abundant sulphide mineral on the planet. Naturally, sulphide minerals have developed under anaerobic conditions found deep underground. Once exposed to aerobic conditions during mining activities, sulphide minerals (including FeS_2) are oxidised, this oxidation process is accelerated in the presence of iron- and sulphur-oxidising microorganisms (Sanchez-Andrea et al., 2014) which catalyse the regeneration of the leach agents Fe^{3+} and H^+.

2.1.2 Environmental impact

As an example of the environmental impact of ARD in South Africa, mining of pyritic coal in the Witbank coalfield of South Africa has led to uncontrolled ARD seepage into the surrounding areas with extensive contamination of ground and surface water (McCarthy, 2011)). The negative environmental impact associated with ARD discharge may vary widely depending on the climate, geomorphology, nature and distribution of ARD generating deposits as well as their relationship to acid neutralising minerals and the associated severity of pH change (McCarthy, 2011). The impact of ARD has contributed to the destruction of both terrestrial and aquatic ecosystems, food chains and ultimately the loss of biodiversity (Feris & Kotze, 2014).

Traditionally the treatment of mine water has focused on pH neutralisation and the removal of heavy metals. Less attention has been placed on the mitigation of dissolved sulphate levels due to their lower environmental risk and regulatory standards when compared to those for acidity and dissolved metals (Arnold et al., 2016). However, regulatory agencies have become increasingly concerned over elevated sulphate concentrations. In some regions, industrial effluents have discharge limits as low as 10 mg/L, although typically this ranges between 250 and 1 000 mg/L (Table 2.1).

Table 2.1: Sulphate concentration discharge levels based on different country guidelines compared to the drinking water standard defined by the World Health Organisation (WHO) (adapted from Arnold et al., 2016)

Authority	Sulphate concentration (mg/L)
South Africa	200 – 600
USA	10 – 500
Canada	500
Finland	2000
Australia	1 000
World Health Organisation	250 (drinking water standard)

Sulphate-rich waste streams are not only produced by mining operations but also as effluents from a variety of industrial operations, including galvanic processing, paper and pulp manufacturing, petrochemical industries, paint and chemical manufacturing, food processing (molasses, oil and seafood), pharmaceutical industries as well as the manufacturing of batteries and chemicals, as summarised in Table 2.2 (Lens et al., 1998; Brahmacharimayum et al., 2019). These industrial effluents may contain a high concentration of sulphate, ranging from 100 to >20000 mg/L. Sulphate concentrations in sewage are typically less than 500 mg/L. The composition of ARD, from leaching of sulphidic minerals, varies significantly, depending on site, environmental conditions, mineralogy and extent of oxidation (Brahmacharimayum et al., 2019) and result in the release of sulphate and hydrogen sulphide contaminated wastewater streams into the environment.

Table 2.2: Industries producing sulphate-rich wastewaters summarising key processes responsible for elevating high sulphate concentration and typical concentration range associated with each industry (adapted from Brahmacharimayum et al., 2019).

Wastewater source	Process from which sulphate-rich effluent is generated	Sulphate (mg/L)	References
Mining	Sulphide mineral oxidation	100 – 20 000	Bai et al. (2013); Banks et al. (1997)
Tannery industry	De-liming, pickling, tanning, re-tanning, dyeing, greasing	2500 - 3000	Galiana-Aleixandre et al (2011)
Chemical industry	Washing of sulfonation reaction products in presence of sulfuric acid	180 000 - 284 000	Sarti and Zaiat (2011)
Sewage	-	<500	Brahmacharimayum et al. (2019)
Drug industry	-	500 - 600	Rao et al. (2007)
TNT (trinitrotoluene) manufacturing process	-	5400	Lens et al. (2010)
Electroplating industry	-	2000	Song et al.(1998)
Galvanic industry	-	200 - 50 000	Tichy et al. (2010)
Citric acid	-	2500 - 4500	Colleran et al.(1995)
Flue gas scrubbing	-	1000 - 2000	Dijkman, (1995)
Alcohol production	-	2900	Lens et al. (2010)
Sea food processing	Wastewaters originating from mussel, tuna, and octopus cooking manufacturing	2100 - 2700	Mendez et al. (1995)
Textile industry	Fish-meal production wastewaters	2690	Kabdasli et al.(1995)
Pulp & paper industry	Thermomechanical pulping	200 - 700	Habets & de Vegt, (1991)
Molasses fermentation	-	2500 - 3450	Carrondo et al. (1983)

The acceptable sulphate limit for taste is <250 mg/L, while international water discharge legislation allows for a sulphate content that ranges between 250-500 mg/L (WHO, 2004). More than 600 mg/L of sulphate is known to cause disturbances in the human gastrointestinal tract often leading to symptoms of diarrhoea, nausea and dehydration. The sulphate concentrations in ARD generally far exceed the permissible discharge levels for human consumption. The excessive release of sulphate, if left unchecked, can lead to pollution of important freshwater resources (surface and ground) and arable agricultural land. This will ultimately have devastating consequence on the water and food security as well as the rich biodiversity of the country.

The effects of ARD are cumulative and present a concerning problem for decades after mining activity have ceased, owing to its ongoing generation. Therefore, the development and implementation of an economically sustainable method for the remediation of ARD pollution is critical in order to preserve the environment, freshwater resource, agricultural land and to ensure human safety (McCarthy, 2011). While it is preferable to prevent the formation of ARD in the first place, once underway, ARD generation can persist for 10s or even 100s of years. This highlights the importance of remediation approaches for legacy ARD sites.

2.1.3 ARD management strategies

2.1.3.1 Preventative control of ARD formation

Ideally, in order to address ARD, source control measures that minimise and prevent the formation of contaminated waters should be implemented (Johnson & Hallberg, 2005). There have been numerous efforts that have focussed on the predictive characterisation and prevention of ARD formation. Although this may be the most preferable solution to address the problem, it may not be a practical or feasible approach in cases where abandoned mines have already begun to decant large quantities of contaminated water. Most prevention strategies, also known as 'source control', are based on the fundamental principle that the formation of ARD is primarily mediated by the exposure of sulphidic minerals to oxygen and water (accelerated by microbial activity) and, therefore, by excluding these factors it may be possible to prevent or minimise ARD formation (Johnson & Hallberg, 2005). Approaches include backfilling, flooding and sealing of abandoned deep mines where dissolved oxygen is consumed by microbial activity and the replenishment of oxygen is hindered by sealing (Johnson & Hallberg, 2005). Other approaches have investigated the role played by iron and sulphur oxidising bacteria in catalysing the generation of ARD, which has led to the use of biocides for the inhibition of their activity within mineral tailings and spoils. However, the application of biocides varies in effectiveness, requires regular application of chemicals and only provides a short-term control to the ARD problem (Johnson & Hallberg, 2005).

2.1.3.2 Migration control

Due to the practical challenges associated with preventing ARD formation at the source, remediation of ARD effluents prior to discharge is required in some cases to reduce its negative impact on receiving water bodies and the surrounding environment (Johnson & Hallberg, 2005). Ideally, the remediation of ARD should neutralise the acidity, decrease sulphate concentration and remove or recover heavy metal contamination (Gopi Kiran et al., 2017). In addition, due to the time-frame of ARD formation and discharge, treatment should be economically sustainable in the long term. Strategies to achieve these objectives can be further subdivided into active and passive treatment processes. A review of active and passive treatment technologies applied in the treatment of ARD is provided in Sections 2.1.5 and 2.1.6.

2.1.4 Sources of ARD

The successful implementation of any remediation strategy is highly dependent on the chemical nature and source of ARD. ARD is characterised by the volume of the effluent, concentration and type of contaminants as well as the pH of the water (Gazea et al., 1996). There are two major sources of ARD that can be distinguished in South Africa. The first is associated with the groundwater rebound from abandoned underground mine workings, primarily from the gold mining impacted basins of the Witwatersrand, Gauteng Province. ARD originating from underground basins are generally characterised by high volumes (several 100 Ml/day) of heavily impacted water containing high sulphate and heavy metal concentration (Rose, 2013). The most appropriate management strategy is to pump and treat, using conventional active processes such as the high density sludge (HDS) process, followed by reverse osmosis (RO). These are costly, require constant addition of alkaline chemicals and do not address the issue of sulphate salinity adequately, however, sustainable alternatives are not yet available to deal with such high volumes (Arnold et al., 2016). As a result, the application of RO for treating ARD is still the most effective.

The second type of ARD originates from diffuse sources leaching from waste rock dumps, spoil heaps and open pits. In South Africa, much ARD generation is associated with the coal industry predominantly within the Mpumalanga province (Figure 2.1). The volume of discharge is significantly lower than that from the underground basins but may vary substantially depending on the site. It has been reported that the long term impact of ARD from diffuse sources, especially from the coal mining industry in South Africa, is likely to affect a far greater area and may persist for a long period of time, given the number of potential sites and the unique combination of climate, geography, distribution, and scale of the deposits. Despite the high risk potential, it has received far less attention from the media, government and mining companies (McCarthy, 2011). Considering the extent of existing and planned mining operations within the region, the management of these sources is of critical importance (Figure 2.1).

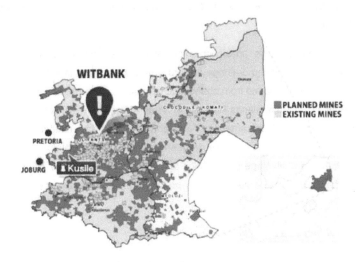

Figure 2.1: Geographic landscape of the coal mining industry in South Africa predominantly within the Mpumalanga province illustrating the extent of planned and existing mines. The vast number of sites will have a significant impact on the surrounding environments with a long term potential risk for the generation and pollution of ARD (Baillie, n.d.).

2.1.5 Active treatment technologies

2.1.5.1 Physicochemical treatment

Active chemical treatment processes using chemical neutralising agents are most widely used for the mitigation of acidic effluents (Johnson, 2000; Shoeran et al., 2010). On addition of an alkaline chemical reagent to ARD contaminated wastewater, increased pH, acceleration in the rate of chemical oxidation of ferrous iron, and precipitation of many metals present in solution as carbonates and hydroxides results. The level at which sulphate is reduced are controlled by the solubility of gypsum which, depending on the ionic strength of the solution, ranges from 1500 to 2000 mg/L. The High Density Sludge (HDS) process is a two stage active process that is based on lime neutralisation, commonly used in the commercial-scale treatment of ARD. This process provides an advantage over conventional chemical neutralising processes in that it provides substantial reduction in sludge volume as well as increases sludge stability (chemically and physically). The HDS process is most widely associated with the refinery and mining industries, and currently it is being used at the Grootvlei Gold Mine on the East Rand. The drawbacks implicated with the HDS process include the special disposal of the resulting sludge, does not significantly reduce salinity levels associated with ARD and has a high chemical consumption suggesting that it is not a sustainable long term solution for ARD remediation. However, it may serve as an effective pre-treatment process prior to sulphate removal (INAP, 2014).

South Africa currently employs physicochemical active treatment systems to treat the voluminous quantities of acidified drainage effluent waters, particularly in the Witwatersrand basin due to ground water rebound from abandoned gold mines. The eMalahleni Mine Water Reclamation Plant in the Mpumalanga Province of South Africa is the result of a joint venture between mining companies Anglo American Thermal Coal and BECSA (Hutton et al., 2009). This treatment facility makes use of active dewatering, followed by oxidation, neutralisation, metal precipitation and multi-stage ultrafiltration and reverse osmosis. The plant produces drinkable water and pure gypsum sludge as a by-product of the ARD treatment (Bezuidenhout et al., 2009).

2.1.5.2 Biological Sulphate Reduction (Bioreactors)

Active systems are generally abiotic based on physicochemical methods. However, there are a number of active treatment technologies that rely on the activity of biological processes. The use of bioreactors is one approach to actively reduce sulphate concentration through exploiting reductive biological processes under defined operating conditions. Biological sulphate removal has the potential to be economically and environmentally favourable treatment option in comparison to current physicochemical processes (Ayangbenro et al., 2018; Nielsen et al., 2018). The major advantages of biological sulphate reduction are: 1) Both sulphate and trace metals can be reduced to low concentration levels, 2) Minimal sludge production 3) Capital costs are relatively low and operating costs can be reduced by the use of inexpensive carbon and electron donor sources, and 4) Trace metals can be selectively recovered, as metal sulphide precipitates, and sold for additional value (INAP, 2014).

Sulphate reducing bioreactors are ideally operated under strict anoxic conditions and rely on the activity of a consortium of specialised microorganisms called sulphate reducing bacteria (SRB) (Sanchez-Andrea et al., 2014). These microorganisms are either autotrophic or heterotrophic organisms and reduce sulphate to sulphide via assimilatory and/or dissimilatory processes (Barton & Fauque, 2009). Biological sulphate reduction can be catalysed by a phylogenetically diverse group of bacteria and some taxonomic genera within archaea, where organic substrate acts as an electron donor while sulphate acts as an electron acceptor (Muyzer & Stams, 2008). These microorganisms are metabolically versatile and can degrade a range of electron donors including ethanol, hydrocarbons, volatile fatty acids, primary sewage sludge and lignocellulosic materials (Liamleam & Annachhatre, 2007; Muyzer and Stams, 2008). However, this is highly dependent on availability and cost of the substrate (Hao et al., 2014). During biological sulphate reduction treating ARD, sulphate is reduced to sulphide in the presence of a suitable electron donor (Reaction 2.2). The process involves the reduction in acidity where the strong acid (H_2SO_4) is transformed into a weaker acid (H_2S) while alkalinity is produced in the form of bicarbonate (HCO_3) (Johnson & Hallberg, 2005).

$$2CH_2O + SO_4^{2-} \rightarrow 2HCO_3^- + H_2S \qquad \text{(Reaction 2.2)}$$

The sulphide generated can be co-precipitated with heavy metals in solution to form stable metal sulphides (Reaction 2.3) (Johnson & Hallberg, 2005). Under the correct pH conditions metal sulphides can be selectively precipitated and recovered and are less soluble than their hydroxide equivalents.

$$Me^{2+} + H_2S \rightarrow MeS + 2H^+ \qquad \text{(Reaction 2.3)}$$

(Me^{2+} is a cationic metal such as Cd^{2+}, Cu^{2+}, Fe^{2+}, Mn^{2+}, Ni^{2+}, and Zn^{2+}).

The application of SRB to treat mine impacted water has been successfully demonstrated at industrial scale. The most recognised being the two patented technologies Biosulphide™ by BioteQ Environmental Technologies Inc., Canada, and Thiopaq™, by Paques, The Netherlands (Ayangbenro et al., 2018). The Thiopaq™ technology incorporates biological sulphate reduction in a gas-lift bioreactor. The Thiopaq™ system consists of two primary stages: 1) An anaerobic stage in which sulphate is reduced to sulphide; and 2) An aerobic stage in which the sulphide, produced in step 1, is oxidised to elemental sulphur. In practice, many variants of the Thiopaq™ technology exist having been tailored to a host of applications treating sulphate- and sulphide-rich industrial waste streams, including the budel Zink (Budelco) refinery (Netherlands) and Kennecott Utah Copper mine (USA) (Hussain et al., 2016).

2.1.6 Passive treatment technologies

Passive treatments refer to the use of natural or constructed wetland ecosystems and are advantageous in that they require very little to no external additions or maintenance (cost) once established (Kaksonen & Puhakka, 2007; Gopi Kiran et al., 2017). These characteristics distinguishes passive from active processes which are highly dependent on maintenance and control of operational conditions, including temperature, pH, and pressure, as well as energy input and the addition of chemicals or substrates (Johnson & Hallberg, 2005). A range of passive treatment options exist and have been applied in different environments, many occurring in various configurations in industry. Well established passive treatments, often used in ARD remediation, include aerobic and anaerobic wetlands and compost bioreactors as well as anoxic limestone drains (ALD).

Passive treatment technologies have advantages over active treatments in that they do not require regular human intervention, operation or maintenance (Johnson & Hallberg, 2005; Gopi Kiran et al., 2017). The construction of passive treatment systems generally incorporates the use of natural materials and promotes the growth of natural vegetation (Skousen et al., 2017). Passive treatment systems are based on the use of gravity flow for water movement instead of active mechanical pumping (Neculita et al., 2008). Ideally a passive system functions without electrical power and can operate for long periods of time (> 5 years). There are several chemical, physical and biological processes that contribute to the amelioration of water quality in passive treatment. These may include adsorption and exchange by soil, plants and other biological materials; metal uptake into live roots and plants; abiotic or microbially-catalysed metal oxidation and hydrolysis reactions in aerobic zones; and microbially-mediated reduction processes in anaerobic zones (Kaksonen & Puhakka, 2007).

A well described method for neutralisation of acidic water is through direct contact with limestone within ALDs (INAP, 2014). Based on its chemical properties, limestone dissolves to deliver calcium and bicarbonate alkalinity; the latter neutralises acidity and buffers pH. The solubility of limestone is dependent on temperature, pH and CO_2 (Johnson & Hallberg, 2005). Mine waters that exhibit a net alkaline characteristic may be treated passively via constructed aerobic (oxidising) wetlands. The system incorporates the abiotic oxidation of ferrous iron and the hydrolysis of the ferric iron produced, resulting in a net acid-generating reaction. Natural vegetation such as macrophytes and Typha are planted for aesthetic reasons to regulate water flow (prevent channelling) and to filter accumulating ferric precipitate (ochre) (Kaksonen & Puhakka, 2007). Additionally, they provide surface area for precipitation of solid phase ferric iron compounds and are also capable of absorption/uptake of heavy metals (Gazea et al., 1996).

In contrast to aerobic wetlands, compost bioreactors make use of anaerobic reactions to mitigate ARD (Johnson & Hallberg, 2005). These systems are enclosed entirely below ground level and do not support vegetation. These systems depend on microbially catalysed reactions that generate net alkalinity and biogenic sulphide. They can be used in the treatment of mine waters that exhibit net acidic and high metal concentrations such as ARD originating from abandoned metal mines. The reductive reactions that occur within compost bioreactors are dependent on electron donors derived from organic material (compost) (Gopi Kiran et al., 2017). The choice of organic material varies based on local availability and effectiveness. Generally, the composts used consist of a mixture of relatively biodegradable materials (e.g. mushroom compost, horse or cow manure) with more recalcitrant materials (e.g. straw and sawdust) (Skousen et al., 2017). In compost bioreactor systems, sulphate and iron reducing bacteria (SRB and FRB) are generally considered to play the major roles in ARD remediation (Johnson & Hallberg, 2005).

2.1.7 Summary of treatment technologies

In summary of Sections 2.1.5 and 2.1.6, the successful remediation strategy of ARD is dependent on accomplishing several objectives, namely the treatment should neutralise acidity, decrease sulphate concentration and remove or recover heavy metals. Furthermore, the ARD treatment adopted should be economically sustainable due to the long term nature of the ARD problem (Gopi Kiran et al., 2017).

Active processes typically include mechanical operations highly dependent on maintenance and control of operational conditions (temperature, pH, pressure) as well as the addition of alkaline chemicals or substrates (Johnson & Hallberg, 2005). These processes are characterised by faster kinetics and enhance control when compared to passive treatments. However, many active systems are unsustainable both from an environmental and economic perspective (Johnson & Hallberg, 2005; Rose, 2013). Alternatively, passive treatments require very little or no external additional maintenance (cost) once installed. However, these systems are dependent on kinetically slower sub-processes and therefore require longer hydraulic retention times (HRTs) and larger areas to obtain effective treatment. In addition, the less defined operating conditions reduce the level of control and predictability (Neculita et al., 2007, Sheoran et al., 2010).

In recent years there has been an increased development of passive bioreactor systems that require periodic active management, such as carbon source addition and/or temperature control, to sustain desired conditions and process performance. Under these conditions, these treatment technologies are referred to as semi-passive (Nielsen et al., 2018). These systems provide an attractive approach over conventional active and passive treatments, with lower capital and operational costs as well as better process control and predictability (Harrison et al., 2014; Nielsen et al., 2018).

Based on this information, when implementing a remedial strategy, careful consideration must be taken based on the advantages and disadvantages associated with a given treatment technology (Table 2.3). Mine location, climate, water characteristics, available utilities and infrastructure, footprint, and disposal areas all preclude a "one-size fits all" solution (Arnold et al., 2016). Active treatments are preferred for treating ARD characterised by high volume, low pH and high metal loading, while passive or semi-passive treatments are typically favoured for treating lower volume and less aggressive discharge of longevity. Therefore, due to the nature of diffuse sources, the application of active treatment is not economically viable, particularly in remote areas where minimal infrastructure is available and contaminated streams are located over large distances. Consequently, biological sulphate reduction has been identified as a promising approach for addressing low volume mine-impacted water through the application of passive and semi-passive treatment (Harrison et al., 2014).

Table 2.3: Overview of ARD treatment technologies categorised as either active or passive, based upon a summary of system performance as well as associated advantages and disadvantages (adapted from INAP, 2014).

	Active Processes			Passive Processes	
	Limestone/ lime	RO	Bioreactor	ALD	Wetland
Pre-treatment	No	Yes	Yes	Yes	Yes
Feed water SO_4^{2-}	3000 mg/L	4920 mg/L	8342 mg/L	3034 mg/L	1700 mg/L
Product water SO_4^{2-}	1219 mg/L	113 mg/L	198 mg/L	1352 mg/L	1540 mg/L
SO_4^{2-} reduction rate	-	-	12-30 g/L.day	-	0.3-197 mg/L.day
Sludge production	Low-moderate	Low	Low-moderate	No	No
Maintenance	Low	High	Moderate	Low	Low
Operating costs	Moderate	High	Moderate	Unknown (low)	Unknown (low)
Advantages	Metal removal Low cost	Water quality	Metal removal	Gypsum product Metal removal	Metal removal Passive treatment
Disadvantages	SO_4^{2-} removal Sludge product	Scaling Membrane lifecycle	Cost of carbon source	Design requirement	SO_4^{2-} reduction

For the purpose of the current research, the subsequent sections of the literature review focus on the potential application of biological treatment approaches for treating sulphate-rich wastewater streams. The sections cover key concepts and develop the rationale for conducting the current work.

2.2 Biological sulphur cycle – treatment options

Sulphur conversions between different forms of reduced and oxidised sulphur compounds involve the metabolism of several specialised microbial communities (e.g. sulphate reducing bacteria and sulphur oxidising bacteria) (Sheoran et al., 2010). These microorganisms possess unique physiological and metabolic traits; novel microorganisms isolated from extreme conditions (pH, temperature and salinity) are reported regularly. These microorganisms are well described throughout literature and have been exploited for industrial application in pollution control of sulphate-rich and sulphide-rich waste streams. Depending on the desired treatment (sulphate reduction or sulphide oxidation), specific microorganisms associated with conversion of sulphur can be cultivated within bioreactors with high efficiency. Although these treatments are aimed at pollution control, new research have focused on the development of a circular economy with a focus on sulphur, metal and water recovery and reuse (Lens et al., 2003).

The enhancement of sulphate reducing processes has become a major focus in ARD treatment in recent years. The advantages of biological treatment over conventional physicochemical treatment are driven by its potential for sustainability, low cost and minimal waste production. Furthermore, depending on the application, these processes can be performed at low temperature with minimal maintenance. An important parameter to the successful operation of biological processes is to understand the biocatalytic reactions that regulate sulphate concentration within these systems. The sulphur cycle is an important biogeochemical cycle and comprises a collection of processes by which sulphur interchanges to and from minerals and living systems (Sheoran et al., 2010; Gopi Kiran et al., 2017). The sulphur cycle consists of four critical steps; 1) Mineralisation of organic sulphur into inorganic forms such as hydrogen sulphide, sulphide minerals and elemental sulphur; 2) Oxidation of hydrogen sulphide, and elemental sulphur to sulphate; 3) Reduction of sulphate to sulphide; and 4) Incorporation of sulphide and organic compounds into metal containing derivatives (Sheoran et al., 2010).

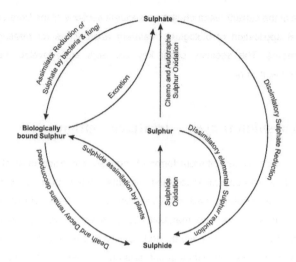

Figure 2.2: The biological sulphur cycle illustrating the intricate network of sulphur transformation that are performed by plants and microorganisms within natural environments such a marine sediments, thermal vents and soil. Many of these reactions are performed by autotrophic and heterotrophic organisms under a range of environmental conditions (Sheoran et al., 2010).

The microbial communities of sulphate reducing bacteria (SRB) and sulphide oxidising bacteria (SOB) are responsible for the cycling of sulphur compounds. These microbial communities play a pivotal role in the development of sustainable biotechnological applications in order to restore the balance in the sulphur cycle such as in the treatment of sulphate-rich wastewaters (Syed et al., 2006; Muyzer & Stams, 2008).

2.3 Biological sulphate reduction

In this section, SRB applied beneficially for treatment of sulphate-rich wastewater are reviewed. For this purpose, SRB are discussed in terms of biochemical pathways, microbial ecology, metabolic requirements and current application in the bioremediation of ARD effluents. Subsequently, key parameters for successful SRB application are also addressed.

2.3.1 Sulphate reducing bacteria

SRB, including both bacteria and archaea, are a unique and highly diverse group of anaerobic microorganisms that obtain energy through the utilisation of sulphate ions (SO_4^{2-}) as a terminal electron acceptor for metabolism of organic substrates (Muyzer & Stams, 2008). Under anaerobic conditions these microorganisms use sulphate as a terminal electron acceptor and couples the oxidation of the substrate (inorganic or organic compound) to the reduction of sulphate. The energy produced is used by the SRB for cellular growth and metabolic maintenance. Although small amounts of reduced sulphur are used for assimilatory sulphate reduction through the synthesis of essential sulphur-containing cellular components, including amino acids and proteins, large amounts are released as free hydrogen sulphide as a generated waste through dissimilatory sulphate reduction for cellular energy (Postgate, 1984). In the context of ARD, as sulphate is reduced to sulphide, there is a reduction in acidity as a strong acid (H_2SO_4) is transformed into a weaker acid (H_2S), while alkalinity produced in the form of bicarbonate (HCO_3^-) neutralises acidity. These reactions are summarised in Reaction 2.4-2.5 (Oyekola et al., 2012; Nielsen et al., 2018):

$$Electron\ donor + SO_4^{2-} \rightarrow 2HCO_3^- + H_2S \quad \quad \text{(Reaction 2.4)}$$

$$HCO_3^- + H^+ \rightarrow H_2O + CO_2 \quad \quad \text{(Reaction 2.5)}$$

The sulphide may remain in solution, evolve as H_2S gas or be coupled to the precipitation of metal sulphides (Reaction 2.6; Johnson & Hallberg, 2005). These precipitates are less soluble than their hydroxide equivalents allowing lower residual metal concentrations in solution.

$$Me^{2+} + HS^- \rightarrow MeS_{(s)} + H^+ \quad \quad \text{(Reaction 2.6)}$$

There have been several studies that have investigated the selective precipitation and recovery of heavy metals in a range of different reactor configurations (Johnson & Hallberg, 2005; Kaksonen & Puhakka, 2007). A study by Santos & Johnson (2018) demonstrated the removal of transition metals (nickel, cobalt, zinc and copper) from synthetic mine water under low pH conditions (4-5). An attractive feature of biological sulphate reduction is that it effectively addresses all three major toxicological characteristics of ARD.

2.3.1.1 Mechanism of dissimilatory sulphate reduction

Understanding the metabolic reactions and ecological significance in SRB is important toward further development and application of BSR processes. There are two primary pathways for sulphate metabolism namely, assimilatory and dissimilatory sulphate reduction. Since the latter is the key mechanism that drives sulphate reduction in anaerobic wastewater treatment, this review only considers the metabolic pathway associated with dissimilatory sulphate reduction. This relies on sequential catalytic reactions in which the reduction of sulphate is coupled with the oxidation of a simple organic compound (Carbonero et al., 2012). Enzymes involved in dissimilatory sulphate reduction including pyrophosphatase, ATP sulphurulase (*sat/atps*), APS reductase (*apr/aps*) and sulphite reductase (*dsr*) (Figure 2.2).

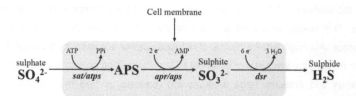

Figure 2.3: Dissimilatory sulphate reduction showing the sequential reaction involved in the reduction of sulphate to sulphide, (e⁻ = electron). Sulphide is actively transported across the cell membrane and activated by ATP forming adenosine phosphosulphate (APS) by ATP sulphurulase (*sat/atps*). APS reductase (*apr/aps*) catalyses the reduction of APS to sulphite which is subsequently reduced to sulphide by sulphite reductase (*dsr*). The generated sulphide is excreted to the environment (adapted from Shen & Buick, 2004).

Dissimilatory sulphate reduction is initiated by the active transport of exogenous sulphate across the bacterial cell membrane into the cell (Figure 2.3). The intracellular sulphate is then reduced in sequential stages to sulphide. Sulphate is highly stable and requires activation prior to subsequent reduction (Brahmacharimayum et al., 2019). ATP sulphurylase catalyses the first reaction to produce the highly activated molecule adenosine phosphosulphate (APS), and release of pyrophosphate (PPi) (Shen & Buick, 2004). ATP hydrolysis in the presence of pyrophosphatase drives the endergonic activation of the sulphate molecule (Barton & Fauque, 2009). APS is converted to AMP and sulphite by the enzymes APS reductase. The sulphite formed is either reduced through a series of sulphur intermediate compounds such as dithionite ($S_2O_4^{2-}$), metabisulphite ($S_2O_5^{2-}$), trithionate ($S_3O_6^{2-}$) and thiosulphate ($S_2O_3^{2-}$), to produce sulphide (Postgate, 1984) or directly through a single step involving a transfer of six electrons in the presence of sulphite reductase. The sulphide generated is then excreted into the environment (Kushkevych, 2016; Rückert, 2016).

2.3.1.2 Microbial ecology

SRB are classified within several different phylogenetic lineages (Muyzer & Stams, 2008). By the year 2009, 60 genera containing over 220 species of SRB were known (Barton & Fauque, 2009). Among the Deltaproteobacteria, SRB are distributed within the orders Desulfovibrionales, Desulfobacterales and Syntrophobacterales. This accounts as the largest group of SRB consisting of 23 genera (Muyzer & Stams, 2008). SRB are also present within a separate thermophilic phylum, known as Thermodesulfobacteria. Additionally, three genera of Archaea, namely *Archaeoglobus, Thermocladium* and *Caldivirga,* are capable of sulphate reduction. SRB are highly abundant throughout nature and have been found to play a functional importance in many ecosystems including polluted environments, cyanobacterial microbial mats, oil fields, marine sediments, purification plants and have been implicated in human disease (Muyzer & Stams, 2008; Barton & Fauque, 2009). Most SRB are characterised by a rod, vibrio or curved cell morphology (Hao et al., 2014).

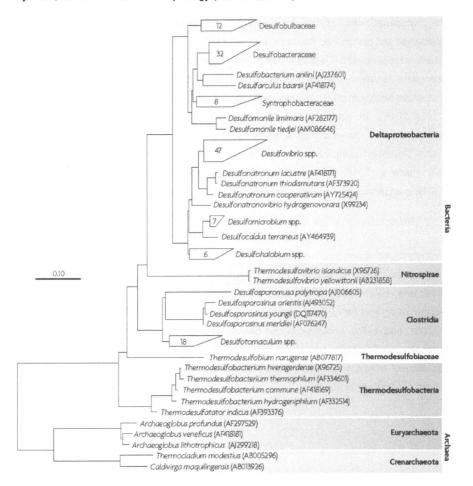

Figure 2.4: Phylogenetic tree based on nearly complete 16S rRNA sequences of described sulphate reducing bacterial species. Sequences were obtained from SILVA small subunit (SSU) rRNA database and the tree was created using ARB software. Note the seven phylogenetic lineages of sulphate reducing bacteria, two in Archaea and five in Bacteria. The number in the collapsed clusters indicates the number of different species within a group. The scale bar indicates 10% sequence difference (Muyzer & Stams, 2008).

SRB are generally classified into two groups based on the metabolic potential to assimilate carbon, those that degrade organic compounds incompletely to acetate and those that completely metabolise organic compounds to carbon dioxide (Muyzer & Stams, 2008). SRB are capable of utilising a range of compounds as electron donor and carbon source including formate, pyruvate, succinate, malate, acetate, propionate, lactate, ethanol, butyrate or a combination of CO_2 and hydrogen (Moosa et al., 2005; Liamleam & Annachhatre, 2007).

The phylogenetic relationships among SRB to metabolise the electron donor have been established and categorised based on genera affiliations. These genera include *Desulfovibrio*, *Desulfomicrobium* and *Desulfobulbus* which represent the major taxonomic lineage of incomplete oxidisers and are the most common SRB genera detected within anaerobic wastewater treatment plants and bioreactor studies (Ito et al., 2002; Hao et al., 2014; Vasquez et al., 2018). In addition, *Desulfobacter*, *Desufarculus*, and *Desulfobacterium* are representative of known complete oxidising SRB genera, while species belonging to the genera *Desulfomaculum* can perform both complete and incomplete oxidation (McDonald, 2007; Hao et al., 2014). These phylogenetic divisions can be used to characterise populations of SRB in anaerobic systems (Muyzer & Stams, 2008).

2.3.2 Bioreactor configurations

A variety of active and passive reactor configurations, have been applied to study anaerobic sulphate reduction or to treat ARD, are summarised in Figure 2.5. These include continuous stirred tank reactors (CSTR) (Moosa et al., 2002; Oyekola et al., 2012), up-flow packed bed reactors (Jong and Parry, 2003; Hessler et al., 2018), membrane reactors (Chuichulcherm *et al.*, 2001) and up-flow anaerobic sludge bed reactors (Sanchez et al., 1997). The selection of bioreactor configuration should consider cost, energy and maintenance requirement, efficiency of mixing and mass transfer, as well as efficient biomass retention to facilitate optimal process performance.

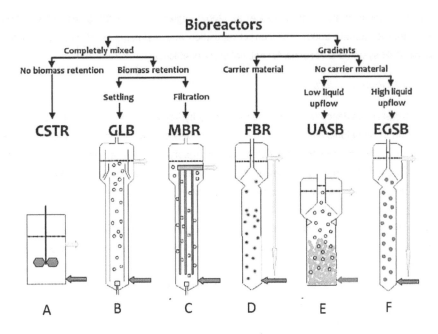

Figure 2.5: Reactor types used for sulphate reduction showing: A) continuous stirred tank reactor (CSTR), B) gas-lift bioreactor (GLB) with internal loop, C) submerged membrane bioreactor (MBR), D) fluidised bed reactor (FBR), E) upflow anaerobic granular sludge bed (UASB) bioreactor, and F) expanded granular sludge bed (EGSB) reactor configurations. These systems can be further categorised by their mixing regime, ability to retain biomass, and use of carrier matrix (adapted from Sanchez-Andrea et al., 2014).

Two main operational configurations have been broadly used for sulphidogenic treatment of ARD: 1) Two-stage reactors where sulphate reduction occurs in one reactor and the produced sulphide is recirculated to another reactor for metallic sulphide precipitation; and 2) A one-stage reactor configuration where sulphate reduction and metal sulphide precipitation occur in a single operational unit (Sanchez-Andrea et al., 2014). The one-stage reactor configuration is preferred for their simple design and reduced capital and operational costs. However, downstream separation and recovery of the metal sulphide can be problematic (Sanchez-Andrea et al., 2014). In most cases multiple reactor units are applied in series to enhance sulphate reduction and metal precipitation (Kaksonen & Puhakka, 2007).

2.3.3 Effect of operational parameters

Numerous studies have investigated the effects of various parameters on the BSR performance such as sulphate concentration, temperature, pH, electron donor availability and type, inhibitory metal and sulphide concentration, as well as the use of different solid support matrices (Elliott et al., 1998; Moosa et al., 2002; Utgikar et al., 2003; Moosa et al., 2005; Baskaran & Nemati, 2006). The correct regulation and maintenance of these parameters within BSR systems is essential for optimal process efficiency.

2.3.3.1 Hydraulic residence time

The hydraulic residence time (HRT) influences the hydraulic conditions within the reactor, the retention of planktonic organisms and the contact between the incoming waste stream and the microorganisms responsible for catalysing the reactions that govern the process. Hence it is an important operating parameter for establishing optimal conditions within different sulphate reducing reactor configurations (Vasquez et al., 2016; Vasquez et al., 2018) and enhancing efficiency (Kaksonen et al., 2004; Greben & Maree, 2000; Dvorak et al., 1992). A short HRT may not allow adequate time for SRB activity to neutralise acidity and precipitate metals, and in the absence of biomass retention may result in cell wash-out. Alternatively, a longer HRT may dictate the depletion of available carbon source or sulphate for SRB activity (Dvorak et al., 1992). Previous studies have revealed that as HRT is decreased (flow rate increase), volumetric sulphate reduction rate (VSRR) increased up until a threshold point after which a further decrease in HRT resulted in considerable loss in system performance. This was attributed to uncontrolled cell washout, proliferation of competitive microorganisms and reaction kinetic constraints associated with BSR (Greben & Maree, 2000; Moosa. et al., 2002; Oyekola et al., 2012).

2.3.3.2 pH

The operating pH of a sulphate reducing bioreactor is a critical parameter for maintaining optimal microbial activity. Most known SRB have been reported as neutrophilic and grow optimally in the pH range of 7.5-8 (Brahmacharimayum et al., 2019). Typically, SRB are inhibited at acidic (<6) and very alkaline (>9) pH ranges. Studies have reported on an optimal operating pH range between 5 and 8. Outside this range, the rate of microbial sulphate reduction significantly declines.

Ideally, to ensure optimal biological sulphate reduction performance a highly acidic waste stream would require pre-treatment neutralisation to limit SRB inhibition. BSR is typically conducted at neutral pH. However, several studies have detected the growth of acidophilic SRB in natural environments growing at pH<3. Several studies have successfully performed BSR treating ARD under acidic conditions (Kolmert and Johnson, 2001; Elliott et al., 1998; Johnson, 1995). A study by Elliott et al. (1998) investigated the effects of acidic conditions on SRB activity. The study evaluated a range of pH conditions in a porous up-flow bioreactor. The system achieved 38% sulphate conversion at pH of 3.25 and 14.4% at pH of 3.0. SRB have been isolated from acidic environments where they exist within microhabitats of suitable pH conditions (Elliot et al., 1998). These microorganisms buffer their surrounding environment by consuming hydrogen as an electron acceptor. A study by Santos & Johnson (2018) demonstrated effective sulphate reduction in an up-flow biofilm bioreactor operated at a low pH of 4-5. The study revealed that the system was dominated by acidophilic SRB namely *Peptococcaceae* strain CEB3 and *Desulfosporosinus acidurans*.

2.3.3.3 Redox potential

For optimal performance, SRB require an anaerobic and reduced microenvironment with a redox potential (Eh) lower than -100 mV (Postgate, 1984). However, the presence of SRB in environments characterised by a positive Eh value has been reported in literature (Neculita et al., 2007). In many of these cases the Eh measurements were collected at the outlet of the bioreactors and did not reflect the microenvironment in which the SRB was present, such as in complex biofilm structures and pores or pockets of organic matter (Sheoran et al., 2010). SRB have traditionally been regarded as strictly anaerobic and are adversely affected when exposed to oxygen (Muyzer & Stams, 2008). However, there has been an increase identification of oxygen tolerant SRB and their survival under oxic conditions (Bade et al., 2000; Sass et al., 2002; Ramel et al., 2015). Members of the *Desulfovibrio*, *Desulfomicrobium*, *Desulfobulbus* genera, most of which are incomplete oxidisers, have been isolated from oxic habitats where they are exposed to oxygen stress (Sass et al., 2002). It has been reported that some of these aerotolerant strains reduce oxygen by re-oxidising sulphide to sulphate, which can subsequently serve as an electron acceptor. This is considered a potential coping mechanism to reduce the toxic effects of oxygen exposure and to maintain anoxic conditions (Cypionka, 2000). A study by Sass et al. (2002) investigated the cheomotaxis behaviour in SRB related to the genera *Desulfovibrio*, *Desulfomicrobium* and *Desulfobulbus*. The study was conducted in oxygen-sulphide counter gradient agar tubes and revealed that SRB move away from high to low oxygen concentrations chemotactically, as a defence strategy. While some SRB are irreversibly inactivated by low oxygen concentrations, others survive aeration even though sulphate reduction is suppressed (Muyzer & Stams, 2008). Cypionka (2000) described SRB that are capable of oxygen respiration to form ATP. Dolla et al. (2006) reported the presence of a superoxide dismutase in some SRB, involved in the molecular strategy to survive in aerobic environment (Dolla et al., 2006).

2.3.3.4 Temperature

SRB can be classified as mesophiles, moderate thermophiles and extreme thermophiles based on their optimum growth temperature (Sheoran et al., 2010). Operating temperatures affects microbial growth, kinetics of organic substrate decomposition, as well as hydrogen sulphide solubility (Sheoran et al., 2010). Under psychrophilic conditions, biogenic alkalinity is hardly produced because of low activity and incomplete oxidation of the electron donor (lactate or ethanol) to acetate. At low temperatures sulphate reduction is kinetically slower, while at the higher temperatures, chemical and enzymatic reaction rates increase (Greben & Maree, 2000). Moosa et al. (2002) reported that sulphate reduction rate increased with increasing temperature from 20 – 35°C in a CSTR, employing a mesophilic SRB culture. Optimum temperature is dependent on the microbial consortium present.

SRB thrive over a wide range of temperatures, exhibiting high flexibility to temperature fluctuations and can generally tolerate temperatures from -5 to 75°C (Postgate, 1984). Studies have reported SRB grow at temperatures as low as 5°C, while some have observed spore-forming thermophilic SRB species grow well at temperature as high as 65 to 80°C (Sheoran et al., 2010). BSR has been recorded at temperatures as high as 100°C (Jeathon et al., 2002; Amend & Teske, 2005). Most SRB are predominantly mesophilic, metabolising optimally at a temperature of 25 to 40°C (Hao et al., 1996; Sheoran et al., 2010)).

Temperature is a critical parameter on the operation of biological sulphate reducing systems particularly where temperature is not regulated within the system. Under these conditions, the key to success of a BSR process is its ability to perform at lower temperatures as well as its resilience to seasonal fluctuation in temperature (Sato et al., 2017).

2.3.3.5 Solid support matrix

The performance of a sulphate reducing system is highly dependent on the microbial biofilm formation on a solid support matrix and its regeneration capacity (Gopi Kiran et al., 2017). The use of solid support matrices in BSR systems facilitates the attachment of microorganisms and enhances biomass retention. This facilitates the decoupling of the biomass retention time and HRT, allowing operation of the system at a high dilution rates, while maintaining high biomass concentration and enhanced reaction rates (Baskaran & Nemati, 2006). SRB do not granulate readily, hence require a solid support on which they can establish microenvironments within biofilms for their survival in the presence of extreme conditions such as high oxygen concentrations and low pH (Lyew & Sheppard, 1997; Gopi Kiran et al., 2017). When SRB have access to a porous surface, higher sulphate reduction rates are observed compared to a free-cell suspension (Glombitza, 2001). The choice of support material is a determining factor in the selection of the active microbial community within a reactor. Solid supports are selected for incorporation into specific applications, hence numerous studies have reported the immobilisation of SRB on different support matrices (Silva et al., 2006). It is essential that the choice of support material within field-bioreactors maintain a balance between surface area and pore size without significantly compromising reactor volume (Neculita et al., 2008; Tsukamoto et al., 2004).

2.3.3.6 COD/Sulphate ratio

An important factor in anaerobic treatment of sulphate-rich wastewaters is the competitive interaction between sulphate reducing bacteria (SRB), methane producing archaea (MPA) and fermentative microorganisms. The COD:sulphate ratio regulates microbial competition. SRB predominate when sulphate is in excess, while under limiting sulphate concentration MPA and fermentative microorganisms dominate (Raskin et al., 1996), based on their relative affinities for the substrate. SRB have a higher affinity for acetate and hydrogen and outcompete MPA at low substrate concentrations (Koizumi et al., 2003). At a COD:sulphate ratio below 0.67 g/g (stoichiometric ratio for complete oxidation), sulphate reduction is favoured over methane production (Oyekola, 2008). Since the treatment of ARD requires the supplementation of organics (carbon source), it is important to regulate organic loading ratio such that complete sulphate removal is achieved whilst minimal residual COD is released within the effluent stream.

2.3.3.7 Inhibitory compounds

Biological sulphate reduction is inhibited when exposed to toxic concentrations of sulphide and heavy metals. In addition, the presence of high concentrations of anions such as sulphate and acetate has also been implicated in inhibiting BSR. The inhibitory action of these compounds is largely dictated by the pH conditions. This has major implications on the efficiency of BSR systems and will be discussed in this section.

Sulphate toxicity: The inhibitory effect of residual sulphate on BSR in a lactate-fed sulphidogenic system was attributed to its effect on the operating pH and redox potential (White & Gadd 1996). An increasing residual sulphate concentration increased redox potential and reduced pH, selecting for a non-SRB community. SRB are known to thrive under low negative redox potentials predominantly within anoxic environments (Postgate, 1984; White & Gadd 1996; Oyekola et al., 2010).

Sulphide and Acetate toxicity: Sulphide (H_2S) and acetate inhibit SRB of which the degree of inhibition is pH-dependent (Reis et al., 1992). The inhibition of SRB activity by sulphide is widely reported (O'Flaherty el al., 1998; Moosa & Harrison, 2006). H_2S has a pKa of 7 at 30°C, therefore, at a neutral pH, sulphide speciation exists as equal amounts of H_2S and HS^- (Figure 2.6). Slight variations in the pH range between 6 and 8 affect the concentration of undissociated sulphide significantly. When pH is dropped to 5, 99% of the sulphide exists as H_2S. The undissociated state (H_2S) is recognised as the inhibitory compound because of its ease in traversing the cell membrane. The inhibitory action of sulphide on microorganism is due to its ability to interact with iron present in cytochrome enzymes or any other metal containing compounds (Sanchez-Andrea et al., 2014). The formation of undissociated sulphide (H_2S) species is favoured by low pH and temperature (Reis et al., 1992). An investigation by Moosa et al. (2006) demonstrated in an acetate-fed SRB culture, in a CSTR maintained at pH of 7.8, that there was a maximum sulphide concentration (1.25 g/L total soluble sulphide) beyond which a significant reduction in sulphate conversion occurred.

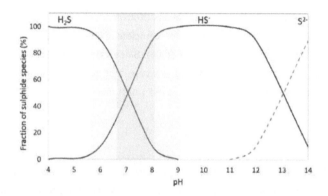

Figure 2.6: Relationship between sulphide speciation at different pH values in an aqueous solution at 30°C. The pH range and optimum of neutrophilic sulphate reducing bacteria are shown in light and dark grey areas, respectively (adapted from Moosa and Harrison, 2006)

Previous studies of acetate-fed sulphidogenic reactors have shown that 50% growth rate inhibition for a mixed consortium of SRB occurred at dissociated sulphide (HS^-) concentrations between 0.4 and 1.04 g/L (total sulphide = 0.57 to 1.11 g/l) as the pH was varied between pH 7.2 and 8.5 (Visser, 1995; O'Flaherty et al., 1998). Acetate in its dissociated form is also known to inhibit SRB activity at low pH values (≤6) (Reis et al., 1992). Similar to sulphide inhibition, it is driven by pH with the unprotonated form able to diffuse across the cytoplasmic membrane and acts as an uncoupler (Baronofsky et al., 1984). The sensitivity of SRB to sulphide and acetate toxicity is also dependent on the bacterial species (Kaksonen & Puhakka, 2007). Several studies have reported that acetate-utilising SRB are more susceptible to sulphide inhibition than compared to incomplete oxidisers (lactate and propionate-utilising SRB). Maillacheruvu & Parkin (1996) studied acetate-oxidising, propionate-oxidising and hydogenotrophic SRB and reported that acetate-oxidising SRB were more sensitive to H_2S concentrations.

Heavy metal toxicity: Metal cations and oxyanions are detrimental to SRB at high concentrations; these are a major toxicological constituent in ARD waste streams. The inhibitory and toxic impact on microbial communities include denaturing proteins, deactivating enzymes and competing with essential cations (Mazidji et al., 1992). Immobilised cells have a greater tolerance to high concentrations of toxic compounds than freely suspended cells (Keweloh et al., 1989). Ideally, heavy metals should be removed prior to biotreatment to prevent adverse impact on SRB population which may lead to significant reduction in process performance (Utgikar et al., 2002).

2.3.3.8 Microbial consortia

Although operating parameters are critical for optimal reactor performance, BSR systems are microbiologically driven. Therefore, the source and selection of the inoculum, survival of the microbial communities, their dynamic evolution, activity and growth are crucial for bioremediation (Gopi Kiran et al., 2017).

One of the biggest challenges facing most biotechnological processes is the lack of knowledge of the active microbial communities responsible for the observed performance. Sulphate reducing bioreactors have been customarily treated as "black boxes" without developing a understanding of the microorganisms that drive the biochemical reactions of the process (Dar et al., 2007). Since the operation of BSR systems is highly dependent on the microbial consortium and its activity, a better understanding of the community dynamics is expected to improve process design and performance (Sheoran et al., 2010). Traditional assessment of microbial ecology and functional potential of the microorganisms within complex environments, such as sulphate reduction bioprocesses, was traditionally based upon culture-dependent approaches which are limited based on overall coverage of the communities.

More recently, the advancement in high-throughput sequencing technologies has facilitated a remarkable expansion in our knowledge of uncultured bacteria and the study of complex microbial communities. In the development of biological treatment, it is becoming increasingly important to understand the microbial ecology dynamics within the bioreactors. The study of the microbial communities and their response to fluctuations to operating conditions has potential to provide a link to the observed performance which can allow optimal parameters to be defined. This is expected to assist in predicting their response(s) to cultivation in bioprocess systems and, thereby, improve the kinetics of sulphate removal from the wastewater streams as well as the robustness of the BSR process.

2.3.4 Biological sulphate reduction treatment technologies

The application of BSR offers an attractive approach to achieving sustainable treatment of sulphate-rich industrial wastewaters. Several developmental studies, based on sulphate reduction have been conducted on laboratory simulated (synthetic) and raw wastewater effluents contaminated with a range of pollutants, particularly containing high concentration of sulphate. Most of these studies were performed at laboratory bench-scale, with limited data available on commercial-scale applications. Though many of these studies show promising results, most are either not economically feasible or failed at demonstration-scale (Hussain et al., 2016). Only three patented technological applications, based on biological sulphate reduction, have been developed and operated at pilot-, demonstration-, and full-scale plants. These include the Thiopaq™ process by Paques (Buisman et al., 2007), Biosulphide® by BioteQ (Ashe et al., 2008) and the Rhodes BioSURE® process (Rose, 2013). However, despite the extensive research on BSR systems, their widespread application remains limited due to several challenges experienced at commercial scale. These major challenges will be explored in Section 2.4.

2.4 Challenges facing biological sulphate reduction treatment

From a technical perspective and review of literature, a study by Harrison et al. (2014) identified three major challenges that would need to be overcome in order to make sulphate reduction technologies more applicable, economically feasible and attractive as a treatment approach. These included 1) the provision of a cost-effective electron donor, 2) the enhancement of reaction kinetics, and 3) improved management of the generated sulphide. These challenges will be discussed in further detail.

2.4.1 Selection of a suitable electron donor

ARD is typically characterised by a low organic carbon content <10 mg/L) and requires the supplementation of an electron donor to facilitate biological sulphate reduction activity (Kolmert & Johnson, 2001). The choice of substrate is governed by several criteria (1) the ability of SRB to assimilate the substrate 2) the suitability of the substrate for the particular application (active vs. passive, reactor configuration, etc.), (3) the amount of sulphate to be reduced and the cost of substrate per unit of sulphide produced 4) local availability 5) potential secondary pollution generated from incomplete degradation of the substrate (Kaksonen & Puhakka, 2007). The type of carbon/electron donor used in BSR can be categorised into two broad groups namely, direct/simple organic substrate or indirect/complex organic substrate. The sequential degradation pathway of these sources for sulphate reduction is presented in Figure 2.7. Since SRB are characterised by their ability to degrade organic carbon, the selection of electron donor has significant implications on the active microbial community within a given bioreactor system and can affect the degree of performance.

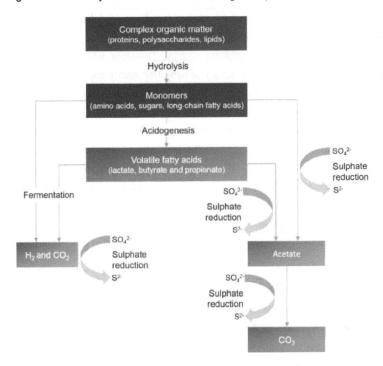

Figure 2.7: Sequential pattern of microbial degradation of complex organic matter in anoxic environment in the presence of sulphate. Macromolecules, such as proteins, polysaccharides, and lipids are hydrolysed by hydrolytic bacteria. Subsequently, the monomers (amino acids, sugars, fatty acids) are fermented by fermentative bacteria during acidogenesis resulting in a range of fermentation products such as acetate, propionate, butyrate, lactate as well as hydrogen and CO_2. In the presence of sulphate, SRB couple the consumption of these products towards sulphate reduction (adapted from Muyzer & Stams, 2008).

2.4.1.1 Direct/simple organic substrate

Direct/simple organic substrates can be defined as easily degradable fraction of organic matter such as low molecular weight compounds with simple structures (Sheoran et al., 2010). These carbon sources do not require decomposition by other microorganisms prior to utilisation by SRB, such substrates include: organic acids (e.g., acetate, butyrate, propionate and lactate); sugars (e.g., sucrose, fructose and glucose); alcohols (e.g., ethanol and methanol) (Liamleam

& Annachhatre, 2007). Direct substrates are preferable in active treatments, it provides better predictability and provides consistent performance. Major drawbacks are associated with the cost and availability of simple substrates.

2.4.1.2 Indirect/complex organic substrate

Indirect/complex organic substrates are those requiring decomposition by a group of microorganisms to provide SRB with easily degradable carbon (Sheoran et al., 2010). SRB are not capable of degrading biopolymers directly. Therefore, when a complex polymeric substrate is used, SRB rely on the activity of hydrolytic and fermentative bacteria. Several studies have investigated the potential of various complex substrates for achieving sulphate reduction. These substrates include lignocellulosic waste (Mooruth, 2013), sewage sludge (Rose, 2013) micro-algal digestate (Harrison et al., 2014) and rice bran (Sato et al., 2017). The sustained release of organic acids from these substrates is an essential parameter when assessing the economic feasibility of a complex carbon source to support SRB over an extended period. A study by Mooruth, (2013), determined that the rate limiting step in the performance of a degrading packed bed reactor, using lignocellulosic material for sulphate reduction, was the hydrolysis and release of soluble organic carbon that can be assimilated by the SRB community.

Depending on the application, a solid, liquid or gaseous substrate can be preferential. For passive treatment, solid complex substrates comprising of plant or waste material are often required to allow passive operation without active pumping of the substrate. However, complex substrates are characterised by a limited lifetime and often require frequent replenishment or will need to be replaced overtime. In active bioreactor configurations, liquid and gaseous substrates are preferred for continuous and consistent process operation providing better control, predictability and stable performance (Sheoran et al., 2010).

An important parameter in selecting a suitable substrate besides cost and availability, is its effectiveness for sulphate reduction. The chemical oxygen demand (COD)/sulphate ratio can be defined as the interaction of SRB with the carbon source available and the electron donor (Gopi Kiran et al., 2017). The minimum COD/sulphate ratio of 0.67 is considered the ideal stoichiometric proportion required for complete sulphate reduction and degradation of the organic substrate. The applied COD/sulphate ratio can vary depending on the type and source of carbon and can range between 0.7 – 1.5. For simple carbon sources the optimum COD/sulphate ratio applied ranges between 0.55 and 0.84 while for organic waste products (activated sludge and municipal compost) the ratio can vary between 1.6 and 5 (Greben & Maree, 2000; Gopi Kiran et al., 2017). The selection of an optimal COD/sulphate ratio is a key parameter in assuring the efficiency of any system designed for treating sulphate. It is important to consider that an effective complex carbon source successfully identified and tested at laboratory-scale may not necessarily be applicable at larger scale. This is largely determined by is availability, sourcing requirements and cost.

2.4.1.3 Lactate and acetate as a carbon source

Lactate based on energy and biomass production is a superior electron donor for SRB activity compared to alternative sources such as acetate, propionate and ethanol (Nagpal et al., 2000). It also facilitates the growth of a diverse range of SRB species. For this reason, lactate has been regarded as the model substrate for studying sulphate reducing activity. It has been applied in numerous kinetic studies for modelling sulphate reduction in continuously stirred tank reactors (Oyekola et al., 2012; Bertolino et al., 2012) as well as a range of different reactor configurations and batch biokinetic tests. Despite the many advantages of lactate in BSR systems, its application has been restricted to laboratory bench-scale studies due to its cost and availability.

Lactate is particularly beneficial for evaluating novel bioreactors systems where the outcome of the study is not constrained by the electron donor to achieve effective sulphate reduction. It has also been implicated in the rapid start-up of BSR systems when supplemented during initial colonisation and biofilm formation (Celis et al., 2013).

Table 2.4: Sulphidogenic degradation reaction of different electron donors and the respective Gibbs free energy change (adapted from Bertelino et al., 2012).

Chemical reaction	$\Delta G°$ (kJ)	
$2\ Lactate + 3\ SO_4^{2-} \rightarrow 6\ HCO_3^- + 3\ HS^- + H^+$	-225.3	Reaction 2.7
$2\ Lactate^- + SO_4^{2+} \rightarrow HS^- + 2\ Acetate^- + 2\ HCO_3^- + H^+$	-160.1	Reaction 2.8
$2\ Ethanol + SO_4^{2-} \rightarrow 2\ Acetate^- + HS^- + 2\ H_2O + H^+$	-66.4	Reaction 2.9
$Acetate^- + SO_4^{2-} \rightarrow HS^- + 2\ HCO_3^-$	-47.8	Reaction 2.10

After lactate, ethanol and acetate are favoured, which have also been extensively studied. While ethanol has been successfully applied at industrial scale (Thiopaq™ process), its widespread application is constraint by local availability and cost. Acetate serves as an important intermediate in the degradation of complex substrates and is generated during hydrolysis and fermentation (Figure 2.7). In addition, the accumulation of acetate as a result of incomplete oxidation of substrates, such as lactate (Reaction 2.8) and ethanol (Reaction 2.9), is the rate limiting step within many high-rate sulphate reducing systems (Kaksonen & Puhakka, 2007). The major drawback associated with acetate can be attributed to the slow growth rate of complete oxidising SRB. Incomplete oxidising associated with lactate metabolism have a doubling time between 3-10 h while complete oxidiser associated with acetate metabolism have a doubling time between 16-20 h (Postgate, 1984; Celis et al., 2013). Thermodynamically, SRB obtain more Gibbs free energy from the incomplete oxidation of substrates (lactate = -160.1 kJ and ethanol = -66.4 kJ) than from the complete oxidation of acetate= -47.8 kJ/mol (Bertelino et al., 2012; Celis et al., 2013).

2.4.2 Biological sulphate reduction kinetics

The second major challenge associated with BSR systems is the reaction kinetics which are governed by several parameters that span metabolic requirements to operational parameters (Section 2.3.3). As previously discussed, the correct regulation and control of these conditions is therefore important to ensure optimal process performance. The feasibility of a wastewater treatment process, besides capital and operational cost, is highly dependent on the quality and rate of treatment. This is largely dictated by the reaction kinetics of the system.

Biological sulphate reduction can be accomplished with freely suspended cells or immobilised cells. In free cell suspended BSR systems, a low dilution rate is required in order to prevent cell wash-out. The wash-out of critical SRB species occurs when the dilution rate exceeds that of the specific growth rate. Since SRB are generally characterised as slow growing microorganisms, in the absence of sufficient biomass retention, BSR processes are governed by long residence times and slow kinetics.

The use of solid support matrices is an important parameter in BSR systems and was briefly introduced in Section 2.3.3.6. The immobilisation of biomass onto a support matrix is preferred in BSR systems in order to decouple the biomass retention time from the hydraulic residence time, allowing operation at high flow rates without significant cell washout. The enhanced biomass retention in immobilised cell bioreactors improves sulphate reduction kinetics (Baskaran & Nemanti, 2006).

The performance of sulphidogenic bioreactor is highly dependent on the microbial biofilm formation and its regeneration capacity (Gopi Kiran et al., 2017). Besides overcoming kinetic constraints observed in freely suspended cell systems, the immobilised cells encapsulated within an EPS matrix tolerate higher concentrations of toxic compounds than freely suspended cells. The outer layers protect the inner layers of the biofilm from exposure to inhibitory concentrations due to mass transfer resistance. In addition, these studies have shown enhance resistance to metals as a result of EPS that protect cells by binding heavy metals and retarding their diffusion within the biofilm (Kaksonen & Puhakka, 2007).

Several solid support matrices have been evaluated for sulphate reduction. These include: polyurethane foam, vegetal carbon, low density polyethylene, alumina based ceramics, sand, glass beads and carbon microfibers (Jong & Parry, 2003; Silva et al., 2006; Baskaran & Nemati, 2006; van Hille et al., 2015; Hessler et al., 2018). The type of reactor configuration (freely suspended cells or immobilisation) and support matrix can affect the ability of SRB to compete for certain substrates as studies have shown preferential immobilisation of non-SRB (Hessler et al., 2018).

A key consideration when selecting a solid support matrix is to ensure minimal compromise on the working volume capacity of the system. Therefore, important factors that governs the effectiveness of a support matrix is whether the material is chemically inert and will not have an adverse (toxic) effect on the microorganisms, the pore size and total surface area, as well as the practical application within the reactor configuration.

2.4.3 Sulphide management

The third major challenged faced by BSR systems is the generation of sulphide during anaerobic treatment which represents one of the bottlenecks associated with the application of these processes and is considered a "secondary pollution" (Pokorna & Zabranska, 2015). Ideally, during the treatment of ARD the generated sulphide is co-precipitated with heavy metals to form stable metal sulphides that can be removed from solution. However, depending on the source of ARD, when there is a deficit in heavy metal concentrations and BSR is highly effective the excess sulphide generated often requires further management. In order to ensure the sustainability of BSR as a long term solution to ARD remediation, the use of an appropriate management strategy of hydrogen sulphide is critical. One approach that has gained a lot of interest in recent years is the potential partial oxidation of sulphide to elemental sulphur. In the following sections a review of sulphide chemistry and application of partial sulphide oxidation for elemental sulphur recovery will be discussed.

2.4.3.1 Sulphide chemistry

In addition to its unpleasant smell, hydrogen sulphide (H_2S) gas is highly toxic (Cai et al., 2017). Upon inhalation, hydrogen sulphide reacts with enzymes in the bloodstream and inhibits cellular respiration resulting in pulmonary paralysis and death. The continuous exposure to hydrogen sulphide concentrations as low as 15-50 ppm results in irritation to mucous membranes and may cause headaches, dizziness and nausea. Higher concentrations of 200-300 ppm may result in respiratory arrest leading to coma and unconsciousness while exposures for more than 30 minutes at concentrations greater that 700 ppm have been fatal (WHO, 2004; Syed et al., 2006)

Sulphide mainly undergoes oxidation by two reactions in the presence of oxygen shown in Reaction 2.11 and 2.12 (Janssen et al., 1999). The reactions represent the oxidation pathways whereby sulphide is converted to sulphate or elemental sulphur under chemical or biological conditions.

$$2HS^- + O_2 \rightarrow 2S^o + 2OH^-$$ (Reaction 2.11)

$$2HS^- + 4O_2 \rightarrow 2SO_4^{2-} + 2H^+$$ (Reaction 2.12)

The Pourbaix diagram (Figure 2.8) provides an indication of the sulphur chemical system in terms of ionic activities as well as thermodynamic forces where equilibrium distribution of dominant sulphur-containing species are represented according to specific Eh (redox) and pH values (Middelburg, 2000).

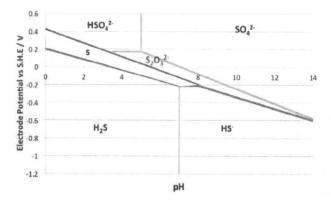

Figure 2.8: Pourbaix diagram Eh = f(pH) representing the predominant sulphur-containing species in equilibrium at different oxygen pressure (redox potential) and acidity (pH) with iron and sodium ions in an aqueous solution (adapted from Middelburg, 2000).

In comparison to the other dominant oxidised forms of sulphur, elemental sulphur is confined to a narrow window of redox potential and pH (Bruser et al., 2000). A study by Nielsen et al. (2005) reported Arrhenius temperature dependency of chemical and biological sulphide oxidation. The rate of chemical and biological sulphide oxidation doubled with temperature increase of 9 and 7°C, respectively. The pH dependence on abiotic and biotic sulphide oxidation is a function of H_2S dissociation with H_2S being oxidised at a lower rate than HS^-. Lewis et al. (2000) suggested that equilibrium thermodynamics impact the major product of sulphide oxidation less than kinetic considerations in biological processes. Conditions present within the bulk chemical phase differ from the intracellular conditions in living systems.

Therefore, further investigation into the microbial kinetics associated with the biological sulphur removal processes will provide insight into process optimisation and conditions that promote partial oxidation of sulphide to elemental sulphur over complete oxidation to sulphate.

2.4.3.2 Sulphide treatment options

The partial oxidation of sulphide to elemental sulphur is an approach that has gained considerable amount of attention. Both physicochemical and biological options have been investigated. Physicochemical processes involved in the removal of sulphide from solution include chemical precipitation and oxidation reactions which usually results in the production of metal sulphide sludge that requires special disposal of. For oxidative reactions, sulphide ions encounter oxygen under controlled redox potential and pH conditions that promote the production of S_0 (elemental sulphur) and hydroxide ions. The produced sulphur with an oxidation state of zero, consisting mainly of cyclic S_8 molecules that aggregate into larger crystals are then separated from solution by flotation or alternative separation techniques (Janssen et al., 1999).

Industrialised physicochemical processes most widely recognized for the removal of sulphide include the Clause process (gas desulfurisation). This multi-step process is used in the petrochemical industry for the recovering of elemental sulphur by stripping gaseous hydrogen sulphide into a glycol or amine solution at high temperature and pressure and subsequently catalytically converting it to elemental sulphur (Elsner et al., 2003). Other processes such as the Stredford process converts sulphide into elemental sulphur in an alkaline solution containing a vanadium catalyst as an oxygen carrier (Kelsall & Thompson, 1993).

The use of the abovementioned physicochemical methods for treating sulphide-rich waste streams is highly effective; however, there are associated disadvantages, such as the high energy input, operational cost and the use of speciality chemicals (catalyst). Due to these requirements, physicochemical treatments are unsuitable for treating ARD (Cai et al., 2017). An alternative approach to achieve sulphide removal with elemental sulphur recovery is the application of biological sulphide oxidation (Harrison et al., 2014).

2.5 Biological sulphide oxidation

The management of hydrogen sulphide is essential for the development of a sustainable, linearised sulphur treatment such that a stable less toxic form of sulphur is removed and recoverable. Strategies reported for the removal of sulphide include metal sulphide precipitation (Johnson, 1995; van Hille et al., 1999; Dvorak et al., 2004), chemical solvent extraction (Johnson, 2000; Janssen et al., 2000) and oxidation to elemental sulphur (Janssen et al., 1999; Molwantwa 2008). The potential application of partial sulphide oxidation to form elemental sulphur within wastewater treatment strategies would not only contribute significantly to the sustainability of the process in managing the sulphide generated during BSR but will also result in the recovery of a sulphur product. This will overcome one of the biggest challenges associated with BSR systems, making it more attractive as a suitable and economically viable approach to treat ARD (Rose, 2002, Harrison et al., 2014).

2.5.1 Sulphur oxidising bacteria

The study of biological sulphide removal under aerobic conditions has been documented since the early 1990. Studies by Buisman et al. (1990) reported the effect of dissolved oxygen and sulphide loading on the performance of biological sulphide oxidation. The study demonstrated that the final product of biological oxidation is highly dependent on the ratio of oxygen to sulphide as represented by Reactions 2.18-2.20 (Guerrero et al., 2016):

$$H_2S + \frac{1}{2}O_2 \rightarrow S^0 + H_2O \qquad \text{(Reaction 2.13)}$$

$$S^0 + H_2O + \frac{3}{2}O_2 \rightarrow SO_4^{2-} + 2H^+ \qquad \text{(Reaction 2.14)}$$

$$H_2S + 2O_2 \rightarrow SO_4^{2-} + 2H^+ \qquad \text{(Reaction 2.15)}$$

These reactions indicate that partial oxidation of sulphide to elemental sulphur proceeds under oxygen-limiting conditions while complete sulphide oxidation toward sulphate occurs when oxygen is in excess.

2.5.1.1 Microbial ecology of sulphur oxidising bacteria

Dissimilatory sulphur metabolism involves the oxidation of reduced inorganic sulphur compounds, such as thiosulphate, polysulphide, elemental sulphur, sulphite and sulphide in the presence of an electron acceptor such as oxygen or nitrate. Biological oxidation of hydrogen sulphide to sulphate is one of the key reactions of the biological sulphur cycle (Figure 2.2). The oxidation of reduced sulphur compounds is performed exclusively by domain archaea and bacteria. Sulphide oxidising bacteria (SOB) are highly diverse, being represented across several genera, and are categorised into two main groups of microorganisms, namely photosynthetic sulphur bacteria and colourless sulphur bacteria.

Photosynthetic sulphur bacteria

Photosynthetic sulphur bacteria, which includes both purple and green sulphur bacteria, use sulphide as an electron donor for photosynthesis, with carbon dioxide as a carbon source in a reaction powered by light. Four major phylogenetic groups of phototrophic sulphur bacteria can be distinguished: a) green sulphur bacteria (GSB), b) purple sulphur bacteria (PSB) and purple non-sulphur bacteria (PNSB), c) Gram-positive Heliobacteria (green non-sulphur bacteria), and d) filamentous and gliding green bacteria (*Chloroflexaceae*) (Barton et al., 2014).

Anoxic phototrophic purple sulphur bacteria are a major group of phylogenetic microorganisms widely distributed through nature. PSB differ from PNSB on both metabolic and phylogenetic grounds, but species of the two groups often coexist within anoxic environments. While PSB are strong photoautotrophs and are capable of limited photoheterotrophy, they are poorly equipped for metabolism and growth in the absence of light (Madigan & Jung, 2008). The growth of these bacteria is highly dependent on light availability, penetration into the water column and wavelength (Syed et al., 2006). In contrast, PNSB, are photoheterotrophs that are capable of photoautotrophy and possess diverse capacities for dark metabolism and growth (Madigan & Jung, 2008).

PSB (>30 genera) consists of a variety of morphological types and belong to the Gammaproteobacteria. The two families of PSB, *Chromatiaceae* and *Extothiorhodospiraceae* produce external and internal sulphur granules, respectively. Genera belonging to PSB include *Allochromatium*, *Thiocapsa*, *Thiospirillum*, *Ectothiorhododospira* and *Halochromatium* (Madigan & Jung, 2008). PSB typically inhabit lakes and hypersaline waters, while PNSB have been isolated from almost every environment, including marine systems, soils, plants, and activated sludge. PNSB (>20 genera) constitute a physiologically versatile group of purple bacteria found within phyla Alphaproteobacteria and Betaproteobacteria. Some well-known PNSB genera include *Rhodobacter*, *Rhodopseudomonas*, *Rhodospirillum*, *Rhodovibrio*, *Rhodoferax*, *Rubrivivax* and *Rhodocyclus* (Madigan & Jung, 2008).

GSB represented by the family *Chlorobiaceae* are obligately anaerobic photoautotrophic bacteria and are found in various aquatic environments particularly in bacterial mats in hot springs, or bottom layers of bacterial mats in intertidal sediments (Barton et al., 2014). Sulphur produced by GSB are excreted extracellularly. Previous studies have reported on the potential application of *Chlorobium limicola,* a photosynthetic sulphur bacterium, in sulphide oxidising bioreactors (Kim et al., 1990). The study reported transformation of 90% of the inlet sulphide to elemental sulphur. However, the use of photosynthetic bacteria is not a viable option in a sulphide removal processes because of the light requirement which complicates reactor design, resulting in high operation cost (Syed et al., 2006).

Colourless sulphur bacteria

The second group of sulphur oxidisers are known as colourless sulphur bacteria for their lack in photopigment. These prokaryotes are generally chemolithotrophs and comprise of an extremely large, heterogeneous collection of bacteria (Syed et al., 2006). Based on comparative analysis of 16S rRNA sequences, the known CSB are assigned into four phylogenetic lineages, three within Bacteria and one within Archaea (Muyzer et al., 2013). Their classification spans across several genera with most CSB affiliated within the phylum Proteobacteria, in particular classes Alphproteobacteria (*Starkeya* and *Thioclava*), Betaproteobacteria (*Thiobacillus* and *Sulfuricella*), Gammaproteobacteria (*Thiomicrospira, Thioalkalimicrobium, Thioalkalivibrio, Thiothrix, Thiohalospira, Thiohalomonas, Halothiobacillus* and *Acidithiobacillaceae*), and Epsilonproteobacteria (*Sulfurimonas, Sulfurovulum* and *Thiovulum*). In addition to Proteobacteria, CSB genera are also found within phylum Firmicutes (*Sulfobacillus*) and Aquificae (*Sulfurihydrogenibium*). Within Archaea, CSB have been classified within the phylum Crenarchaeota belonging to the genera *Sulfurisphaera, Acidanus* and *Sulfolobus* (Muyzer et al., 2013; Barton et al., 2014).

The most abundant and documented group of CSB are in the family *Thiobacilliaceae* in terrestrial environments and in the family *Beggiatoaceae* in aquatic environments. Representative organisms within the domain Archaea are represented by aerobic microorganisms belonging to the Order Sulfolobales and are characterised by thermophiles (extremophiles that require high temperatures to grow) and acidophiles (extremophiles that require low pH to grow) (Pokorna & Zabranska, 2015). CSB are phlylogenetically and physiologically versatile and are found is almost all natural environments where reduced sulphur compounds are available (e.g., in soil sediments, at aerobic/anaerobic interfaces in water, and in volcanic sources such as the hydrothermal vents). CSB are typically found growing at neutral to slightly alkaline pH values, though, acidophilic CSB have been reported in ARD water (Barton et al., 2014).

SOB genera belonging to the family *Thiobacilliaceae* has been widely studied and applied in the treatment of sulphide-rich waste streams (Syed et al., 2006). Although most SOB require oxygen, some (such as *Thiobacillus denitrificans*) can grow anaerobically and utilise nitrate or nitrite as an alternative electron acceptor during denitrification. CSB have been isolated from a variety of environments, such as soda lakes, marine sediments, wastewater treatment plants and thermal springs (Syed et al., 2006). These bacteria are generally aerobic mesophilic microorganisms that grow optimally within a pH and temperature range of 2 to 7 and 20 to 30°C, respectively. Many of these bacteria can grow under various environmental stress conditions and exhibit high conversion rates that make it suitable for industrial application (Buisman et al., 2010).

Other SOB genera that have demonstrated high sulphide oxidation conversion capabilities and potential for application in industrial treatment have included *Pseudomonas putida* (Syed et al., 2006), *Alcaligenes sp.* (Kantachote et al., 2008) *Paracoccus pantotrophus* (Vikromvarasiri et al., 2017b) and *Halothiobacillus neapolitanus* (Vikromvarasiri et al., 2017a).

2.5.1.2 Mechanism of sulphide oxidation

Biological sulphide oxidation processes can be divided into aerobic or anaerobic treatment based on different electron acceptors. Under aerobic treatment, oxygen serves as the electron acceptor while during anaerobic treatment nitrate or nitrite is used as the electron acceptor (Cai et al., 2017). However regardless of the process, the pathway of inorganic sulphide transformation in SOB is similarly regulated. To ensure the recovery of elemental sulphur and to prevent complete oxidation toward sulphate. It is important to regulate oxygen-limiting conditions to favour sulphur recovery.

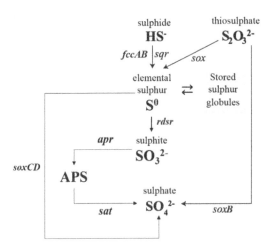

Figure 2.9: Sulphur oxidation pathway showing the major reactions and enzymes responsible for oxidising reduced sulphur compounds such as sulphide (incomplete). *Sqr* and *fccAB* catalyses the oxidation of sulphide leading to the deposition of elemental sulphur, *soxCD* oxidise sulphur directly to sulphate. Alternatively, *rdsr* can oxidise sulphur to sulphite, and subsequently to sulphate by *apr* and *sat* via the intermediate formation of adenosine phosphosulphate (APS). Thiosulphate is oxidised to sulphate via *soxB*.

A variety of enzymes catalysing inorganic sulphur oxidation reactions have been identified and characterised in sulphur oxidising bacteria (SOB). The enzymes include the membrane bound sulphide:quinone oxidoreductase (*sqr*), the periplasmic flavocytochrome c sulphide dehydrogenase (*fccAB*), reversible dissimilatory sulphite reductase (*rdsr*), sulphur oxidising multi-enzyme system (*sox*), adenosine phosphosulphate reductase (*apr*), and ATP sulphurylase (*sat*) (Barton et al., 2014). The sulphide oxidation pathway is presented in Figure 2.9.

The oxidation of sulphide is mediated by either *fccAB* or *sqr*, which produce sulphur predominantly in the form of polysulphide. In most chemolithotrophic SOB belonging to Gammaproteobacteria oxidise thiosulphate to sulphur by the *sox* multienzyme system consisting of *soxXAYZB*, that lacks *soxCD*. From zero-valent sulphur, there are two pathways in which sulphate is formed, i) a direct reaction whereby *soxCD* facilitates a six-electron oxidation and ii) a pathway where sulphur is oxidised to sulphite by *rdsr* gene (reverse of dissimilatory sulphate reduction) and subsequently oxidised to sulphate by *apr* and *sat* via the intermediate formation of APS (Barton et al., 2014; Berben et al., 2017).

2.5.1.3 Bioreactor configurations

The major challenges associated with these processes is the difficulty of maintaining microaerobic conditions through conventional bubbling aeration. In addition, the supply of oxygen requires large amount of energy and may result in striping of hydrogen sulphide from the water which can be a hazardous concern accompanied by an unpleasant odour release and will affect the sulphur recovery efficiency (Cai et al., 2017). The application of biological sulphide oxidation has been studied under different reactor configurations and air supply for optimisation of sulphur recovery including continuous stirred-tank reactors (Buisman et al., 2010), gas-fed batch reactors (Jensen & Webb, 1995), expanded granular sludge bed reactors (Xu et al., 2012) and airlift reactors (Dogan et al., 2012).

2.5.2 Generation of biologically produced sulphur

2.5.2.1 Characteristics of biologically produced sulphur

Biologically produced sulphur is divided into two groups, based on whether they are produced by internal and external sulphur excretion mechanisms (Cai et al., 2017). Internal sulphur is present within the periplasm SOB as invaginations that are often encapsulated by a protein envelope though to be purely structural in function. External sulphur exists as globules that are not enclosed in the cell membrane and is the preferred form for harvesting from wastewater treatment processes (Pokorna & Zabranska, 2015; Cai et al., 2017).

According to previous reports, biological sulphur forms transparent globules which are deposited on the inside or outside of SOB. Biologically produced sulphur has a white or pale colour and a higher refractive index than water (Cai et al., 2017). X-ray measurements of sulphur obtained from sulphide oxidising bioreactors revealed it is partly built up of orthorhombic sulphur crystals (S_8). Orthorhombic sulphur is usually highly insoluble in water (5 g/L), although it can be readily dissolved in a non-polar solvent such as hexadecane. However, solubility tests of biologically produced sulphur particles indicated that they were soluble in water, rather than in hexadecane, indicating that they are hydrophilic. The hydrophilic characteristic of biosulphur was attributed to the presence of amphiphilic compounds covering the hydrophobic S_8 nucleus. A study by Janssen et al. (2001) conducted electrophoretic mobility measurements and flocculation experiments on biosulphur and suggested that the sulphur particles are covered by an extended negatively charged polymeric layer, most likely made up of protein.

2.5.2.2 Recovery of biological produced sulphur

Based on the inherent characteristics of biologically produced sulphur several physical or physicochemical methods have been employed for the recovery of elemental sulphur from water or sludge. Gravity sedimentation of sulphur is undoubtedly the cheapest and technically the most attractive approach (Cai et al., 2017). However, it is reliant on the settling properties of the sulphur particles. In most cases biological sulphur exists in the form of colloid which cannot be precipitated directly. Therefore, this requires acid treatment in order to destabilise the sulphur particles. The sulphur can thereafter be recovered by an inclined plate precipitation method. At a pH of 5 and a surface hydraulic load less than or equal to 0.42 $m^3/m^2.h$ occurring in the sedimentation tank, the separation rate of sulphur remains above 80%.

The use centrifugation methods are a relatively well documented approach which has been widely used in industry. An example of its application is in the Thiopaq™ process. Produced elemental sulphur in the aerobic bioreactor is separated from the aqueous phase in a separator inside the reactor as a slurry of 20% (w/w) solids content. The sulphur slurry is dewatered using continuous decanter centrifuge resulting in a sulphur cake made up of 60-65% dry solids. After centrifugation sulphur purity can reach up to 98%. Alternatively, membrane separation techniques involving the recovery of sulphur by membrane surface filter have also been developed. A study by Camiloti et al. (2016), demonstrated the use of a tubular silicone rubber membrane for sulphur recovery by controlling the air supply and creating oxygen limiting conditions that facilitate partial sulphide oxidation to form elemental sulphur. The results showed that biological sulphur was deposited on the surface of the membrane. By changing the hydrodynamic conditions in the membrane, either by increasing the flow rate or changing its directional flow the sulphur would detach and subsequently be recovered.

Biological sulphide oxidation is an attractive alternative treatment due to its feasibility of operation at ambient temperature and atmospheric pressure with reduced energy cost.

Biological oxidation expresses effective sulphide removal, reduced pollution in the form of secondary products and does not require the use of hazardous chemical reagents (Janssen et al., 2001). This gives rise to less environmental pollution in comparison to conventional physicochemical processes (Buisman et al., 2010). Development of these systems, however, requires careful consideration when selecting the microbial community to convert H_2S to S^0 (Syed et al, 2006). The feasibility of a given sulphide removal process is dependent on the operational cost, maintenance, removal efficiency as well as the ease of separation and recovery of the sulphur product (Syed et al., 2006; Cai et al., 2017).

2.5.3 Biological sulphide oxidation treatment technologies

The study of sulphide oxidation for sulphur recovery has been studied in a variety of reactor configurations. However, like BSR systems its widespread application at industrial scale has been largely limited to few technologies applied under niche environments that enable their feasibility. The most successful commercialised systems are the THIOPAQ™ SULFATEQ™ and THIOPAQ™ O&G processes developed by Paques and Paqell (joint venture between Paques and Shell), respectively (Cline et al., 2003). These processes will be introduced in the current section.

The THIOPAQ (SULFATEQ™) process operated at the Budelco zinc refinery in the Netherlands utilises the sulphide oxidising ability of SOB (Janssen et al., 2001; Cline et al., 2003). The overview of the process is presented in Figure 2.7.

The process treats zinc sulphate-containing process water. Sulphate reduction takes place in full-scale (500 m^3) sulphate-reducing gas-lift reactor. Synthesis gas is used as the electron donor by steam-reforming natural gas. The inlet gas is comprised of approximately 76% hydrogen, 20% carbon dioxide, 3% nitrogen and 1% carbon monoxide. Sulphide generated precipitates zinc which is collected in a settler and reused in the roasting process. Excess sulphide is directed toward the aerobic bioreactor under oxygen-limiting conditions, where partial sulphide oxidation occurs resulting in the recovery of elemental sulphur downstream (Muyzer & Stams, 2008).

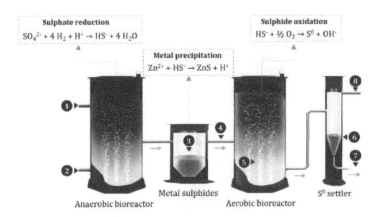

Figure 2.10: Schematic overview of the THIOPAQ™ SULFATEQ™ process to remove sulphate and heavy metals from waste water. 1) Sulphate and metal contaminated water enter the anaerobic bioreactor where sulphate-reducing bacteria reduce sulphate to sulphide using 2) hydrogen (H_2-rich gas) as an electron donor. Subsequently, the sulphide generated is used to 3) precipitate the heavy metals. The 4) excess of sulphide is converted to elemental sulphur by sulphide oxidising bacteria in a 5) aerated bioreactor. The precipitated metal sulphides and elemental sulphur can be 6) separated and 7) recovered while the 8) purified effluent is safe for discharge (adapted from Paques, 2019).

The THIOPAQ™ O&G process has been successfully applied as a desulphurisation treatment of sulphide-rich gas streams at multiple refineries. The process incorporates an alkaline scrubber which absorbs H_2S into solution and subsequently oxidises the sulphide to elemental sulphur within the aerobic bioreactor (Cline et al., 2003).

Most chemical and biological sulphide oxidation (THIOPAQ™) processes are operated actively as a downstream process treating sulphide-rich waste streams generated at chemical refineries. The major drawback of these systems is the high process costs, maintenance and use of specialised equipment. This has significantly hindered its application in the treatment of ARD discharge. However, the discovery of floating sulphur biofilms (FSB) and their potential application in the treatment of ARD wastewater has provided an promising approach to achieving partial sulphide oxidation with the recovery of elemental sulphur, under passive operating conditions (Molwantwa et al., 2010; Mooruth, 2013)). The occurrence of FSB and its potential application for treating sulphide-rich waste streams will be discussed in the following sections.

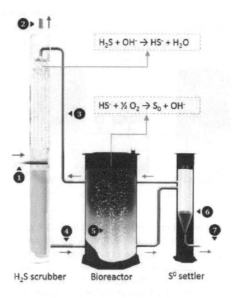

Figure 2.11: Schematic overview of the THIOPAQ O&G process to treat sulphide-rich gas streams. 1) Sulphide-rich gas is absorbed into solution and 2) purified gas released with 3) an alkaline wash solution within the H_2S scrubber system. Subsequently, the 4) sulphide-rich solution is converted to elemental sulphur by sulphide oxidising bacteria in the 5) aerated bioreactor. The generated elemental sulphur is 6) separated downstream within a settler and 7) recovered (adapted from Paques, 2019).

2.6 Floating sulphur biofilm in wastewater treatment

The generation of elemental sulphur through biological sulphide oxidation occurring in several natural environments have been well documented. One biological structure of high biotechnological interest is floating sulphur biofilms (FSB), which were found forming naturally on the surface of highly sulphidic tannery wastewater ponds (Molwantwa, 2010). These sulphur-rich films were also observed adhering onto the glass walls of sulphide-rich reactors at the gas-liquid interface (Oyekola, 2008). These floating films were later identified to consist of a diverse microbial community comprising of sulphide oxidising bacteria (SOB) which are involved in the conversion of sulphide to biological elemental sulphur. Through elemental analysis these biofilms were predominantly comprised of elemental sulphur. This generated further interest into understanding the function and relevance of FSB occurring in sulphide-rich water bodies and its potential application for treating sulphide-rich wastewater (Molwantwa, 2010).

A biofilm is defined as an aggregation of microorganisms in which cells that are often enclosed within a self-produced extracellular polymeric substance (EPS) adhere to each other and/or to a surface or interface (Paytubi et al., 2017). Biofilms that form at the air-liquid interface, are generally referred to as "pellicle". The air-liquid interface serves as a favourable niche environment for the growth of bacteria where nutrients are acquired from the liquid and oxygen from the surrounding environment. There has been a growing interest in the study of pellicle formation, with most studies conducted on pure cultures including *Bacillus subtilis* (Kobayashi, 2007), *Shewanella oneidensis* (Armitano et al., 2013), *Acinobacter* baumannii (Chabane et al., 2014) and *Salmonella enterica* (Paytubi et al., 2017).

The ability to colonise the air-liquid interface to form a floating structure requires high organisation due to the lack of a solid surface for initial attachment (Chabane et al., 2014). The formation of floating biofilms or pellicle begins with the attachment of planktonic cells to the surface. Flagella and pili appendages of microorganisms have shown to play an important role in the migration and attachment of cells (Chabane et al., 2014). After reaching the air-liquid interface, bacteria adapt to the environment and an increase in EPS synthesis is initiated. A homogenous layer forms after which the biofilm matures into a three-dimensional biofilm structure where bacterial cells are covered and connected by an intricate EPS network. A study by Armitano et al. (2013) reported the importance of aerotaxis (movement of motile cells, in the direction corresponding to an increasing gradient of oxygen) in floating biofilm development.

The formation of FSB, though more complex in microbial community dynamics and structure, exhibits similar developmental stages described in pellicle formation of pure cultures at the air-liquid interface. Studies by Molwantwa (2010) and Mooruth (2013), performed an extensive analysis on the formation and structure of FSB. The studies demonstrated the potential application of FSB to achieve partial sulphide oxidation with high elemental sulphur recovery. A conceptual model based on its development and structure is presented in Figure 2.12.

The application of the floating sulphur biofilm for achieving partial sulphide oxidation was evaluated in a linear flow channel reactor (LFCR) (Molwantwa, 2008). The reactor was constructed for the cultivation of the FSB and comprised of two channels with a total surface area of 0.55 m^2 and total volume of 0.022 m^3 (2.5 m x 0.11m x 0.04 m). Baffles were placed along the length (0.5 m apart) of the reactor to control the flow of the water through the reactor (

Figure **2.13**).

The sulphide feed to the LFCR was received from a sulphide generating lignocellulose degrading packed bed reactor (DPBR). The study achieved an average sulphide removal of 65% and sulphur recovery of 56% at a sulphide loading rate between 1309 and 2618 L/m^3/d. The results were promising, demonstrating the potential of the LFCR for treating sulphide-rich wastewater. The LFCR would be modified over time with addition of more channels and redesigned for field demonstration. This led up to the eventual up-scale application of the process as part of a semi-passive treatment, known as the Integrated managed passive (IMPI) process.

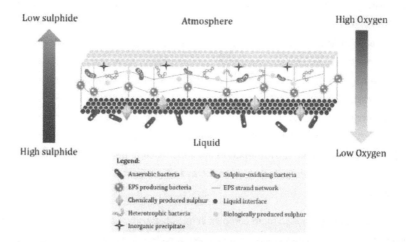

Figure 2.12: Conceptual structural model of the FSB at the liquid-air interface of a sulphide-rich liquid. Steep dissolved oxygen and sulphide gradients, as indicated by the blue and red gradient arrows, respectively, are established at the air-liquid interface. During biofilm formation EPS producers migrate to the air-liquid interface and generate a slime layer which constitutes the framework of the biofilm. Aerobic heterotrophic bacteria and sulphide oxidising bacterial consortia are established within the EPS network. At the correct sulphide/oxygen ratio, partial sulphide oxidation is favoured and an increased activity of SOB results in the accumulation of biological sulphur in the biofilm. Inorganic precipitates are formed by evaporative crystallisation. On the underside of the FSB, as oxygen diffusion across the

biofilm becomes limiting the bulk volume becomes anoxic providing suitable conditions for the proliferation of anaerobic microorganisms. Figure adapted from Molwantwa (2008) and Mooruth (2013).

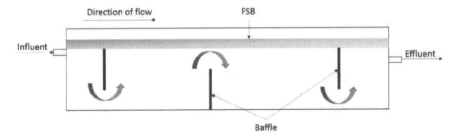

Figure 2.13: Illustration of the linear flow channel reactor showing the placement of baffles and the fluid flow pathway through the length of the channel with the formation of the FSB at the air-liquid interface (adapted from Molwantwa, 2008).

2.7 Application of biological technologies to treat ARD in South Africa

In this Section, biological treatments that have been applied at demonstration scale in South Africa for treating ARD will be reviewed.

2.7.1 Paques THIOPAQ™ process

The Paques THIOPAQ™ process described in Section 2.5.3 was implemented at the Landau Colliery in Witbank to treat ARD at the Navigation site. The plant was commissioned as one of the options to be evaluated by Anglo Coal and Ingwe Collieries (BHP Billiton) for the Emalahleni Mine Water Reclamation plant. The design specifications were for the treatment of 20 ML per day, with a water recovery of greater than 95%. The raw water had a pH of 3.12 and contained sulphate (2,500 mg/L), calcium (536 mg/L), iron (81 mg/L), aluminium (16 mg/L) and manganese (23 mg/L). Target concentrations for product water quality were sulphate (<200 mg/L), calcium (<30 mg/L) and heavy metals at less than 0.15 mg/L. A demonstration scale plant, capable of treating 3 ML/day, was commissioned and operated for several years. The performance of the plant was encouraging, with effluent sulphate concentrations typically below 500 mg/L which was within discharge limit (Table 2.1). The operators did however encounter challenges, particularly with scaling of the heat exchangers. In addition, the licensing fees and the cost of the electron donor (ethanol) counted against the technology.

Despite relatively stable and effective performance the system was eventually decommissioned (Günther and Mey, 2006).

2.7.2 Rhodes BioSURE®

The Rhodes BioSURE process was developed at Rhodes University based on the observation of enhanced degradation of complex organic wastes in sulphate reducing tannery ponds (Rose, 2013). These observations led to the development of an integrated process for treating ARD that relied on algal production to provide the electron donor for sulphate reduction. The system was known as the Integrated Algal Sulphate Reducing Ponding Process for Acid Metal Wastewater Treatment (ASPAM) process. However, though the treatment was an attractive option for treating low volume discharge, an increase in mine closures at the Witwaterstrand basins and the extent of ARD discharge, necessitated the need for a more readily available electron donor to be sourced. This led to the identification of primary sewage sludge (PSS) as a promising alternative. The use of PSS as an electron donor proved effective and prompted the development of the recycling sludge bed reactor (RSBR) (Rose, 2013).

The RSBR concept was developed and scaled up through 2 L, 10 L, 3000 L to 23000 L reactor configurations. In these studies, a multi-compartment baffle reactor was investigated in a second stage unit. In the second reactor the soluble and suspended COD, derived from the hydrolysis of the PSS flocs in the RSBR, provided a readily available electron donor for sulphate reduction. The sludge bed in the upflow chambers of the baffle reactor, provided immobilisation of SRB and entrapment of particulate organics. The control of the COD:sulphate was an important parameter to prevent a shift to methogenic conditions.

The system was successfully tested at pilot (40 m^3/day) and demonstration scale (1.6 ML/day). The encouraging performance of the demonstration plant ultimately led to full scale implementation in 2005 at the ERWAT Ancor sewage treatment works at the Grootvlei Gold Mine located in Eastern Witwatersrand, South Africa, where PSS was sourced. The full scale plant received 10 ML/day of post-HDS effluent from the Grootvlei Gold mine and 2 ML/day of iron-hydroxide sludge (Rose, 2013). The system consisted of eight upflow sludge blanket reactors with external sludge recycling. The process was designed to remove sulphate to levels below 250 mg/L.

The process was successfully operated for several years but was eventually decommissioned. This was primarily due to changes at the PSS utility, rather than failure of the technology. Although the process demonstrated improved sustainability and feasibility for treating ARD, the HDS pre-treatment required large amounts of lime and energy. In addition, the system was reliant on the existing municipal treatment works to supply the electron donor. Its widespread application would be limited based on these requirements. Ultimately, the system would not be applicable for treating lower volume discharge in remote locations.

2.7.3 Integrated Managed Passive (IMPI) process

The disadvantages of active treatment systems and the enhanced sustainability of more passive based systems are highlighted in Section 2.1.7. Due to the nature and persistence of low volume discharge from diffuse sources often located in remote areas, there has been a need for the development of sustainable passive/semi-passive treatment options. These technologies need to be cost effective and require minimal maintenance. The integrated managed passive (IMPI) process, by a South African company (Pulles Howard and de Lange) in collaboration with the Environmental Biotechnology Research Unit (EBRU) at Rhodes University, was developed in order to address these recalcitrant low volume discharges. The technology was a semi-passive process centred on biological sulphate reduction and sulphide oxidation for sulphate removal and elemental sulphur recovery, respectively.

The IMPI process comprised of a patented DPBR, which formed the basis of the treatment. The DPBR relied on the establishment of three distinct microbial communities. The first community was responsible for removing oxygen from the system to ensure the necessary redox environment (-250 to -350 mV). The second community characterised by *Clostridium* species facilitated the degradation of the lignocellulosic material and generation of soluble substrate. The third community, made up of SRB, coupled the oxidation of available carbon to the reduction of sulphate, generating alkalinity in the form of bicarbonate and hydrogen sulphide. The resulting effluent, released from the DPBR, characterised by high sulphide concentration was then fed into a novel sulphide oxidation unit downstream, known as the linear flow channel reactor (LFCR), which facilitated partial sulphide oxidation of sulphide via the formation of a FSB. At full scale the overall configuration of the IMPI process consisted of four operating units in series shown in Figure 2.14.

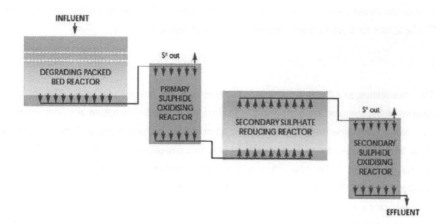

Figure 2.14: Schematic diagram of the IMPI process, comprising of primary and secondary degrading packed bed reactors and sulphide oxidising units. The DPBR were optimised for degradation of lignocellulosic material and VFA release for high biological sulphate reduction while sulphur oxidising units oxidised the generated sulphide toward elemental sulphur via the formation of floating sulphur biofilms. The secondary reactors were predominantly served as polishing steps to treat any residual contaminants (adapted from Pulles & Heath, 2009).

The full scale system was designed to treat 200 m^3 of mine water and was constructed at the Middelburg mine in Mpumalanga, South Africa. The hydrolysis of the lignocellulosic material was identified as the rate limiting step in maintaining efficient system performance. The DPBR was characterised by four phases which included the lag phase (90-150 days), where the microbial communities acclimatise to the environment, a high performance phase (< 8 months), during which high sulphate reduction efficiency is maintained, followed by a crash phase and sustained phase, at which point the hydrolysable lignocellulose becomes depleted and sulphate reduction performance decreases to relatively low rates.

2.8 Current state of research and development

Following the initial demonstration of the IMPI process, the pilot plant faced several challenges, particularly the sulphide oxidation component, which did not perform as expected (van Hille et al., 2011; van Hille and Mooruth, 2014). Research undertaken at the Centre for Bioprocess Engineering Research (CeBER) at the University of Cape Town, highlighted several fundamental problems with the IMPI process and optimised the sulphide oxidation LFCR unit (Mooruth, 2013). The LFCR was redesign and constructed in Perspex which allowed for the hydrodynamics to be assessed. The study by Mooruth, (2013) identified that the original LFCR design resulted in the short-circuiting of the fluid at the base of the reactor which resulted in the poor performance of the process. The LFCR was subsequently re-designed and optimised for sulphide oxidation. The study demonstrated the feasibility of achieving high partial sulphide oxidation and sulphur recovery in a sulphide-fed LFCR. The study reported that a 1 and 2 day HRT was optimal with sulphur recovery ranging as high as 75 to 92%, respectively. Mooruth (2013), also noted that the ratio of oxygen to sulphide supplied to the biofilm played a critical role in the formation of the desired product and that sufficient carbon source was important for effective biofilm formation. The study concluded that the correct regulation and maintenance of these influencing factors was essential for process efficiency.

Following the successfully optimisation of the LFCR reactor a study by van Hille et al. (2015) demonstrated the potential application of carbon microfibers as a solid support matrix for biological sulphate reduction within a closed (anaerobic) LFCR configuration. As a result, the reactor was sealed with a top lid which limited oxygen ingress into the reactor. The carbon microfibres provided a high surface area for biomass retention without significantly reducing the effective reactor volume capacity, commonly associated with many bulky solid support materials. The study achieved a high sulphate reduction conversion that ranged between 85 and 95% at a feed sulphate concentration of 1 g/L. During operation at a dilution rate of 0.083 1/h, while substantial cell washout was observed in a continuously stirred-tank reactor (CSTR), the LFCR maintained a VSRR approximately 20% higher than the CSTR (van Hille et al., 2015). During the study, complete elimination of oxygen was not possible and there was evidence of partial sulphide oxidation and the establishment of a FSB similar to that observed in the dedicated sulphide oxidation reactor (Mooruth, 2013). This suggested that partial sulphide oxidation could be coupled with biological sulphate reduction within a single LFCR configuration.

2.9 Research rationale and motivation

2.9.1 Integration of sulphate reduction and sulphide oxidation in a hybrid LFCR

Biological sulphate reducing systems are well described in literature and have been applied as active systems. Passive SRB systems described are limited to traditional wetlands or DPBR. The main drawback associated with these systems is unpredictable system performance (Zagury et al., 2007). For biological sulphide oxidation systems, studies are confined to active treatment systems. These systems make use of either membrane or gas-lift reactors (Hurse & Keller, 2004; Henshaw & Zhu, 2001). A major drawback is the energy requirement for continuous oxygen supply, light and a large surface area. In addition, ARD treatment systems may be located in remote regions where there is a lack of infrastructure and electricity. Based on this, current passive and active SRB bioremediation systems are not feasible to address the long period of time required for effective treatment of ARD.

The integration of sulphate reduction and partial sulphide oxidation is highly desirable in the treatment of sulphate-rich waste streams as this would effectively linearise the conversion of sulphate to elemental sulphur and overcome one of the major drawbacks associated with many BSR treatments. In addition, the recovery of elemental sulphur as a value from waste, greatly increases the feasibility of the process and can be applied in agriculture (fertiliser) and chemical (sulphuric acid) industry. The benchmark of such an approach incorporating biological sulphate reduction and partial sulphide oxidation has been the THIOPAQ™ process reviewed in Section 2.5.3 and 2.7.1. However, it is not applicable for long term treatment of sulphate-rich ARD emanating from abandoned mines or diffuse sources in remote locations where passive treatments are preferred. Thus, there is still a need for the development of sustainable semi-passive treatment options that provide an effective and sustainable approach for addressing low volume ARD from diffuse sources.

As reviewed in the preceding Section 2.6 and 2.8, there have been several studies that have contributed toward the development of an integrated semi-passive process (Figure 2.15).

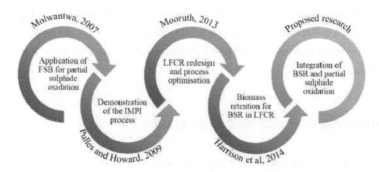

Figure 2.15: Contribution of work conducted by several authors across different organisations and institutes, towards the development of an integrated semi-passive process. These studies were instrumental in the initial development of the LFCR and application of FSB for partial sulphide oxidation. Collectively, major findings and observations from these studies has led to conception of the current proposed research.

The research began since the initial investigation at the Environmental Biotechnology Research Unit (EBRU), at Rhodes University, into the possible role FSB could play in the treatment of sulphide-rich wastewaters based on the observation of its formation on tannery waste ponds (Rein, 2002). This eventually led to the investigation into the application of FSB for sulphide oxidation in a novel LFCR which detailed the structure of the FSB with regards to microbial ecology and mass transport of oxygen through the biofilm (Molwantwa, 2008). The successful lab demonstration resulted in the up-scale of the process as part of the IMPI process developed by Pulles Howard and de Lange and Golder Associates Africa (Pulles & Howard, 2009). However, the performance of the system did not meet to expectation. A detailed study into the fundamental aspects of reactor hydrodynamics, organic loading rate and sulphur speciation as well as operating parameters conducted at the Centre for Bioprocess Engineering, at the University of Cape Town, resulted in the optimisation of the LFCR design (Mooruth, 2013). Further studies demonstrated the operation of a closed LFCR and its potential for achieving high sulphate reduction through enhance biomass retention (Harrison et al., 2014; van Hille et al., 2015). The cumulative contribution of this research conducted by several authors, organisations and institutes forms the foundation and rationale for the current work.

The primary aim of this study is to build on the observation by Harrison et al. (2014) and van Hille et al. (2015), by investigating the potential integration of biological sulphate reduction and partial sulphide oxidation within a single hybrid LFCR unit. Due to the novelty of such a process, a comprehensive study would be required to characterise the biochemical processes that underpin its successful operation. Few technologies exist that incorporate both biological sulphate reduction and partial sulphide oxidation within a single reactor unit. This is mainly due to the contrast in conditions required for each process and therefore is generally performed in separate reactor systems. Understanding the complexity of the hybrid system from a process performance and microbial ecology perspective of both SRB and SOB microbial communities will be important toward the development and application of the process as a sustainable approach for treating low volume sulphate-rich wastewaters, such as ARD.

Against the background of the literature review and in contribution to the ongoing research initiative on BSR and ARD treatment at the Centre for Bioprocess Engineering Research group, UCT, the current work investigates the demonstration and characterisation of a novel semi-passive process for achieving simultaneous biological sulphate reduction and partial sulphide oxidation with the recovery of sulphur under semi-passive conditions within a single reactor unit.

2.10 Research scope

This study provides proof of concept of the one stage integrated bioprocess for ARD treatment to provide an elemental sulphur product and a remediated water stream with reduced salinity and acidity. The study investigates and provides new insight into the operation of the integrated process and links process kinetics and microbial community dynamics to changing operational conditions. This understanding will prove valuable for characterisation and further optimisation of the process.

2.10.1 Research hypotheses

Hypothesis 1

Based on the hydrodynamics of the linear flow channel reactor design and the limited turbulent mixing, it is possible to create discrete anaerobic and aerobic zones within the reactor, such that biological sulphate reduction and biological sulphide oxidation can occur simultaneously, with the formation of a structurally sound floating sulphur biofilm.

Hypothesis 2

Characterisation and optimisation of the kinetics and microbial community dynamics of the semi-passive integrated process as a function of operational parameters (temperature, HRT and sulphate loading) provides insight into the successful operation and management of the novel wastewater treatment system.

Hypothesis 3

The use of a lactate or an acetate feed selects for different microbial community structures with the latter favouring complete oxidation. Both these substrates provide appropriate organic substrates for the sulphate reducing and sulphide oxidising communities to sustain efficient system performance; however, acetate is preferred owing to the improved per mol carbon to sulphate ratio required and the reduced residual COD released.

2.10.2 Research objectives

- Demonstrate "proof of concept" of the integrated channel reactor for simultaneous sulphate reduction and sulphide oxidation.
- Develop an efficient harvesting system to recover elemental sulphur from the floating sulphur biofilm, that minimises the time required for the biofilm to reform.

- Evaluate the overall process performance of the hybrid LFCR, in terms of sulphate reduction efficiency and kinetics, sulphide oxidation efficiency and recovery of elemental sulphur using a defined growth medium.
- Investigate and optimise sulphate reduction and partial sulphide oxidation performance based on the effect of hydraulic residence time, temperature and sulphate loading.
- Investigate and optimise sulphide oxidation efficiency based on biofilm collapse and harvesting regime.
- Evaluate the use of acetate as an alternative electron donor to lactate.
- Investigate the relationship between microbial ecology, operating conditions and process performance, in terms of community composition and relative abundance.

2.10.3 Research strategy

The main objectives of the present study were addressed through sequential phases, presented in Figure 2.16. The research strategy outlines an interconnected approach to comprehensively characterise the process across a range of operational parameters. Broadly, the experimental studies included initial demonstration of the hybrid LFCR process at lab scale, process performance based on a range of operational parameters (HRT, temperature, sulphate loading, electron donor and reactor geometry) and linking microbial community dynamics with system performance. The key objectives to be addressed for each experimental study are presented at the beginning of the respective chapter.

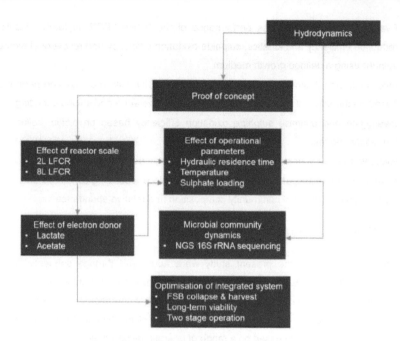

Figure 2.16: Illustration of the research strategy adopted in this work

Chapter 3

Materials & Methods

This Chapter presents the detailed methods used to conduct the research reported in this thesis. This includes a full explanation on the experimental setup and analytical methods. Where an experiment or study(s) employs an alternative approach, an explanation for its use, experimental setup and operation is addressed within the respective chapter(s).

3.1 Experimental setup

3.1.1 Microbial cultures

The SRB mixed microbial community (stock culture), originally sourced from the laboratory of Prof Peter Rose (Department of Microbiology, Biochemistry and Biotechnology) at Rhodes University, South Africa, has been maintained at the University of Cape Town (UCT) over an extended period on modified Postgate B medium (Table 3.1) since 2001. The culture was originally derived from an anaerobic compartment of a facultative pond at the Grahamstown sewage treatment works. Depending on the experiment the medium was made up in 1 L (batch cultures) and 10 L (continuous bioreactor operation) Schott bottles.

Table 3.1: Modified Postgate B medium composition and quantities used in the current study (Oyekola et al., 2012).

Component	Amount
KH_2PO_4	0.46 g/L
NH_4Cl	1.0 g/L
$MgSO_4.7H_2O$	2 g/L
$NaSO_4$	0.3 g/L
yeast extract	1 g/L
sodium citrate	0.3 g/L
60% sodium lactate (w/w)	1.6 ml

The synthetic feed was sterilised by autoclaving at 121°C, 103 kPa for 20 min. The sulphide oxidising bacteria (SOB) consortium was developed at UCT using enrichments from SRB reactors (van Hille and Mooruth, 2013).

3.1.2 Linear flow channel reactor configuration

The linear flow channel reactor (LFCR) applied to this study has been designed for semi-passive operation and requires low maintenance. The bulk volume of the reactor is exposed to the surrounding environment in which the air-liquid interface facilitates the formation of the FSB. While the LFCR has been applied as a sulphide oxidation unit, in the current investigation the LFCR was modified to achieve simultaneous biological sulphate reduction and partial sulphide oxidation with sulphur recovery.

The 2 L LFCR was constructed from Perspex (11 mm thickness) and had internal dimensions of 250 mm (l) x 100 mm (w) x 150 mm (h) (Figure 3.1). The front facing side of the reactor was fitted with nine sampling ports, allowing the bulk reactor volume to be monitored across the length and at different heights. The reactor design was based on the original 25 L LFCR described by Mooruth (2013).

Additionally, the 2 L LFCR configuration was modified and fitted with a plastic strip (10 mm wide) holding carbon microfibers as a microbial support matrix as well as a submerged heat exchanger (4 mm ID) for temperature maintenance and control (

Figure 3.2). A screen, made of plastic mesh fixed to an aluminium frame, was designed to lie 5 mm below the liquid surface to facilitate biofilm capture and harvesting.

3.1.3 Linear flow channel reactor operating conditions

The reactor was operated at a feed sulphate concentration of 1000 mg/L supplemented with lactate as a sole carbon source to maintain a chemical oxygen demand (COD) to sulphate ratio of 0.7. The reactor was operated at 28°C and neutral pH. As illustrated in

Figure 3.2, a peristaltic pump (ISMATEC) was used to pump the feed in continuously from the uppermost inlet-port on the left side of the reactor while the effluent flowed from an equivalent exit-port on the right side of the reactor.

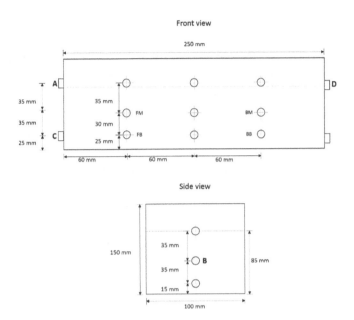

Figure 3.1: Schematic diagram of the 8 L LFCR simulating the dimensions of the pilot scale in aspect ratio showing the front and side view and the location of A) inlet port, B) carbon microfibers, C) heat exchanger, and D) outlet port. FM (front middle), FB (front bottom), BM (back middle), and BB (back bottom) represent the sampling ports provided and used for regular positional sampling during the study.

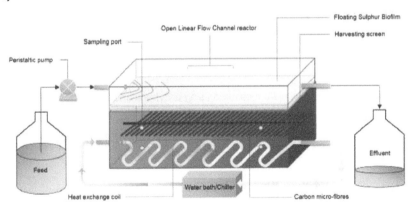

Figure 3.2: Schematic diagram of the hybrid LFCR configuration showing reactor set-up and operation. Sulphate-rich media was fed into the inlet port at a defined flow rate through peristaltic pump. Overflow into the effluent reservoir was governed by position of the exit port. Circulation of heated or chilled water from a temperature-controlled bath through the fitted heat exchanger regulated the operating temperature. The reactor was fitted with bulk liquid sampling ports and carbon fibres for biomass retention. A harvesting screen just below the interface facilitated biofilm collapse and harvesting. The reactor surface was exposed to the atmosphere.

3.1.4 Sampling layout

The six sampling ports were sampled regularly using a 19 gauge epidural needle (8 cm) to monitor the solute concentrations in the reactor fluid across the length and depth of the LFCR. In addition, the length of the needle ensured sampling was taken near the centre of the bulk volume. The sampling locations included the middle and lower sampling ports in the first and third columns labelled as Front Middle (FM), Front Bottom (FB), Back Middle (BM) and Back Bottom (BB), respectively, as shown in Figure 3.1. In addition, samples of reactor overflow were collected as a representative effluent sample. The multiple sampling points across the bulk volume allowed monitoring the distribution of physicochemical parameters, including aqueous species, throughout the LFCR to evaluate system performance. It also provided insight into mixing of the LFCR during operation where the presence of stratification or short-circuiting of the influent stream could be detected. The samples were subjected to a range of solution chemistry and analytical methods, described in Section 3.2, to monitor reactor performance.

3.2 Analytical methods

3.2.1 Chemical reagents

The chemicals and reagents used throughout the duration of the research were of analytical grade sourced from Merck, Accsen Instumental, Kimix, and Thermo Fisher Scientific.

3.2.2 pH and REDOX potential

The pH and redox potential were measured using a Cyberscan 2500 micro pH meter and a Metrohm pH lab 827 redox meter fitted with a Metrohm Redox platinum-ring electrode, respectively. The pH probe was calibrated daily using Accsen Instrumental standard buffering solutions (pH 4.0 and 7.0).

3.2.3 Hydrogen sulphide analysis

Dissolved sulphide was quantified using the colorimetric methylene blue technique (APHA, 2012). The method is based on the ability of hydrogen sulphide and acid-soluble metallic sulphides to convert N, N-dimethyl-p-phenylenediamine directly to methylene blue in the presence of a mild oxidising agent (acidified ferric chloride). The intensity of the methylene blue colour development is directly proportional to the amount of sulphide present in the tested sample. The colorimetric measurement of this intensity through spectrophotometry provides an accurate means to determine the sulphide concentration (Kovooru et al., 2013).

Briefly, an appropriate volume (20 μL) of sample collected from the experimental system(s), as described above, was added to 200 μL of 1% (w/v) zinc acetate immediately after sample collection and made up to 5 mL using dH$_2$O. To this was added 500 μL N,N-dimethyl-p-phenylenediamine hydrochloride solution, followed by 500 μL of ferric chloride solution. This mixture was vortexed for 10 s and allowed to react for 15 min at room temperature before the absorbance was read at 670 nm (A_{670}). The final sulphide concentrations were determined by interpolation from a sulphide standard curve that ranged between 0.2 - 1.0 mg/L. Details of the standard curve and reagent preparation are detailed in Appendix A.1.

3.2.4 Sulphate and thiosulphate analysis

Two methods for determining sulphate concentrations were employed within this study, namely ion chromatography (IC) analysis as well as the barium chloride method (APHA, 2012).

3.2.4.1 Ion chromatography determination

Residual sulphate and thiosulphate concentrations were determined by IC using a Thermo Scientific DIONEX ICS-1600 system equipped with an IonPac AG16 anion column, a 10 μl injection loop and a conductivity detector with suppression. A 22 mM NaOH solution was used as the mobile phase at a flow rate of 1 mL/min, as per manufacturers' recommendations, and analysis was performed using the Chromeleon®7 software package (version no. 7.2.1.5833).

3.2.4.2 Barium chloride method

The barium chloride method, based on the turbidimetric analysis of barium sulphate formation in solution, was also used to quantify the residual sulphate concentration. The method is based on the precipitation of SO_4^{2-} ion in solution with $BaCl_2$ forming barium sulphate ($BaSO_4$) which remains suspended in solution, causing opacity. The concentration of sulphate can be determined turbidimetrically through spectrophotometry (APHA, 2012).

Analysis was conducted on samples prepared following the removal of dissolved sulphide, by addition of 40 μl 10% (w/v) $ZnCl_2$ solution to 2 mL sample to precipitate ZnS. The samples were vortexed vigorously for 5 s before centrifugation (13,000 rpm for 15 min at room temperature), to remove the ZnS precipitate, and the supernatant was filtered. These samples were appropriately diluted into 5 mL dH$_2$O, followed by the addition of 250 μL of conditioning reagent and an excess amount of finely ground $BaCl_2$ to facilitate the precipitation reaction.

The conditioning reagent prevents the formation microcrystalline $BaSO_4$ and stabilises the suspension. This reaction mixture was vortexed for 60 s after which absorbance was measured at a wavelength of 420 nm (A_{420}) using a spectrophotometer (VWR® model: V-1200). Sulphate concentrations were determined from the absorbance readings using a sulphate standard curve ranging from 10 - 50 mg/L. For standard curve and reagent preparation see Appendix A.2.

3.2.5 Sulphur analysis

Elemental sulphur concentration was determined by High Pressure Liquid Chromatography (HPLC; Thermo Scientific System spectraSYSTEM AS3000) using a Discovery HS (C18) octadecylsilane reverse phase column (25 cm x 4.6 m, 5 µm), with detection by absorbance at 263 nm (A_{263}). A 95% methanol solution (HPLC grade) was used as the mobile phase and was operated at a flow rate of 2 mL/min for 10 min. A sample injection volume of 20 µL was selected.

All samples underwent a rigorous sample preparation. Briefly, colloidal sulphur was recovered from reactor and effluent samples (2 mL) by centrifugation (13 000 rpm for 10 min at room temperature). The supernatant was discarded, and the sulphur-containing pellet dissolved in 1 mL chloroform (100%) in a microfuge tube. These samples were vortexed for 1 minute and incubated for 30 minutes at 50°C in a dry block, to facilitate the dissolution of the colloidal sulphur, before being vortexed for another 2 min. The solution was filtered through a 0.22 µm Millex nylon syringe filter into HPLC sample vials (Mooruth, 2013). Samples were immediately analysed by HPLC and quantified relative to a sulphur standard curve, prepared as described above between 0 - 8 mM S_0. The standard curve and standard solution preparation are detailed in Appendix A.3.

The accurate measurement of polysulphides is complex and challenging. A study by Mooruth (2013) developed a technique to quantify polysulphides but was time consuming and extremely hazardous. The study concluded that under the operating conditions similar to the current investigation, the formation of polysulphides was negligible. Due to the low contribution of polysulphide on the overall sulphur balance and challenge associated with quantification, polysulphide concentration was not measured in the current investigation.

3.2.6 Volatile fatty acids

Volatile fatty acids (VFAs) analysis was conducted to quantify the concentration of lactic, acetic and propionic acids in the feed and reactor samples. VFA concentration was determined using high pressure liquid chromatography (HPLC) on a Waters Breeze 2 system equipped with a Bio-Rad organic acid column (Aminex HPX-87H, 30 cm x 7.8 mm, 9 μm) and a UV (210 nm wavelength) detector. Acidified deionised water (0.01 M H_2SO_4) was used as the mobile phase at a flow rate of 0.6 mL/min (Mooruth, 2013; Biorad, 2012). Samples were prepared by appropriately diluting with dH_2O and filtering samples through a 0.22 μm Millex nylon syringe filter into HPLC vials. Standard solutions (0.1 – 0.6 g/L) were prepared by performing serial dilutions of acetate, propionate and lactate stock solutions (10 g/L), respectively, with dH_2O. The standard curve and standard solution preparation are detailed in Appendix A.4.

3.3 Floating sulphur biofilm (FSB)

3.3.1 Biofilm disruption and harvesting

The floating sulphur biofilm (FSB) develops at the air-liquid interface of the bulk fluid within the LFCR. The FSB was collapsed, on a regular basis, by physically disrupting the biofilm with a spatula and allowing the fragments to settle on the mesh-screen positioned just below the surface (termed disruption). Following disruption, the biofilm can re-form at the surface. The sulphur product was recovered by removing the mesh-screen and collecting the accumulated biofilm (termed harvesting). The biofilm was dried at 37°C, weighed and stored for elemental analysis.

3.3.2 Elemental analysis

The CHNS analyser is based on the principle of "Dumas method" which involves the complete and instantaneous oxidation of the samples by flash combustion (≥1800°C). The combustion products are separated by a built in gas chromatographic (GC) column and detected by the thermal conductivity detector (T.C.D) which gives an output signal proportional to the concentration of the individual components (CO_2, NO_2, SO_2, and H_2O) of the mixture.

Elemental analysis of the harvested FSB was determined using an Elementar Vario EL Cube Elemental Analyser, for quantifying carbon, hydrogen, nitrogen and sulphur content (Central Analytical Facility (CAF), Stellenbosch University, South Africa).

3.4 Determining mixing patterns by a dye tracer study

The LFCRs were constructed from clear poly(methyl methacrylate) (Perspex) to allow for easy visualisation of the hydrodynamic mixing patterns. A dye tracer experiment was conducted by filling the LFCR to its standard operating volume with 2 mM sodium hydroxide to which ten drops of pH indicator dye phenolphthalein was added to achieve a uniform pink colour. A solution of 42 mM hydrochloric acid was pumped into the reactor at a pre-determined flow rate, representing the feed rate of interest. When the neutralisation reaction occurred (Reaction 3.1) the liquid within the reactor turned colourless, thus demonstrating the fluid's path (Mooruth, 2013).

$$NaOH + HCl \rightarrow NaCl + H2O \qquad \text{(Reaction 3.1)}$$

The duration of each experiment varied based on how long it took for the entire reactor fluid volume to turn colourless. This was recorded photographically at time intervals appropriate to the mixing time. The mixing study was repeated as a function of reactor scale, residence time and temperature.

3.5 Scanning electron microscopy

Most of the current knowledge of biofilms is due to advances in microscopic image techniques. Scanning electron microscopy (SEM) has been shown to be a suitable tool not only for detailed visualisation of bacterial biofilm morphology and structure, but for following biofilm formation and adhesion onto abiotic surfaces. The application of SEM is highly advantageous due to its high resolution and magnification which enables the observation of the physiological shape of the microorganisms composing the biofilm as well as the spatial organisation (Gomes & Mergulhao, 2017). In the current work SEM was applied to visualise and study the floating sulphur biofilm as well as the attached biomass on the carbon microfibers. In addition, energy-dispersive X-ray spectroscopy (EDS) analysis coupled with SEM was used for elemental analysis on the biofilm to confirm elemental sulphur deposits and composition within the FSB and to characterise the crystalline precipitate within the FSB.

The technique was performed on the FSB and colonised carbon microfibers that were aseptically removed from the LFCRs. The samples were fixed within a 2.5% (w/v) glutaraldehyde in 1X phosphate-buffered saline (PBS) solution for 24 hours at 4°C. PBS (10X) was prepared as follows: 8 g of NaCl, 0.2 g of KCl, 1.44 g of Na_2HPO_4 and 0.24 g of KH_2PO_4 was dissolved in 800 ml. The pH was adjusted to 7.4 and made up to a final volume of 1 L with dH_2O before autoclaving. After glutaraldehyde fixation, the samples were washed twice with 1X PBS solution followed by being dehydrated through an ethanol series, namely 30, 50, 70, 80, 90, 95 and 100% (v/v) ethanol, with a 10 min incubation at each step. Samples were then carefully mounted onto SEM stubs with carbon glue and critical point dried (CPD) using hexamethyldisilazane (HMDS). The dried samples were coated with gold-palladium (60:40) and observed using a FEI NOVA NANOSEM 230 (Botes et al., 2002).

3.6 Microbial community dynamics

One of the biggest challenges facing most biotechnological processes is the lack of knowledge into the active microbial communities. Sulphate reducing bioreactors have been traditionally treated as "black boxes" without any thorough understanding of the microorganisms that drive the biochemical reactions of the process. Since the operation of BSR systems is highly dependent on microbial activity, a better understanding of the community dynamics will help to improve process design and performance (Sheoran et al., 2010).

The study of microbial communities through culture dependent techniques are limited when profiling complex microbial communities. In the past few decades the development and application of culture independent techniques such as denaturing gradient gel electrophoresis (DGGE), terminal restriction fragment length polymorphism (T-RFLP), Fluorescence *in situ* hybridisation (FISH) and Genechips were used as mainstream methods in the study of bacterial communities and diversity (van Hille et al., 2015; Yang et al., 2016)). More recently, the development and advancement of high-throughput sequencing technology have provided a new approach to evaluating microbial communities. Metagenomic methods provided by next-generation sequencing platforms such as Roche 454 and Illumina have facilitated a remarkable expansion of knowledge regarding uncultured bacteria. The technique is primarily based on the phylogenetic sequence analysis of the 16S rRNA amplicon (Yang et al., 2016).

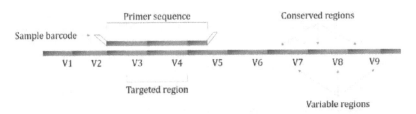

Figure 3.3: Schematic diagram of the 16S rRNA gene illustrating all nine variable and conserved regions. Amplification of the targeted region (V3-V4) of interest using a specific primer set designed with the complementary sequence and labelled with sample barcode.

The 16S rRNA gene sequence was first used in phylogenetic analysis in the early 1980s (Woese, 1980). Its highly conserved regions for the design of universal primers and hypervariable regions to differentiate and identify phylogenetic characteristics of microorganisms, has made it the most widely used marker gene for profiling bacterial communities (Yang et al., 2016). The full-length of the 16S rRNA gene sequence consists of nine hypervariable regions that are separated by nine highly conserved regions (Figure 3.3).

In most studies the targeted 16S rRNA gene sequences are partially complete, using designed primers that amplify specific regions of the gene, due to limited access to sequencing technology and cost. A simplified outline of the procedure is depicted in Figure 3.4. Selected environmental samples are subjected to genomic DNA extraction followed by a PCR amplification of the 16S rRNA gene. The amplicons are subsequently sequenced using an NGS platform. The raw sequences are subsequently aligned, trimmed and filtered through a bioinformatics pipeline (QIIME) (Caporaso et al. (2010). In addition, phylogenetic classification and statistical analysis are performed. Several visualisation tools are used for data analysis.

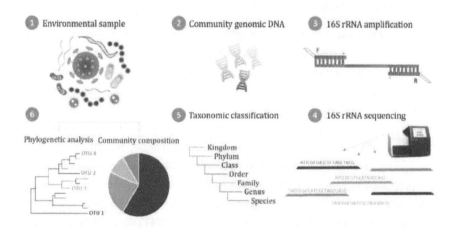

Figure 3.4: Illustration briefly outlining the procedure for 16S rRNA metagenomics sequencing. 1) Environmental sample collection, 2) genomic DNA extraction 3) PCR amplification of the 16S rRNA sequence 4) 16S rRNA sequencing using next generation sequencing platform; the raw sequences are filtered, trimmed and clustered into operational taxonomic unit (OTUs) 5) OTU taxonomic assignment 6) community composition and phylogenetic analysis.

The selective amplification of a target sequence (16S rRNA) in a mixed DNA sequencing has its disadvantages and can be biased. The technique is limited by the amplification of short read lengths, sequencing errors, variation among different regions, difficulties in classification at species level and poor resolution among closely related species (Poretsky et al., 2014). In addition, it cannot be used for absolute quantitative analysis. However, it does provide an extensive summary of microbial diversity and relative abundance within a given sample. The resolution of the phylogenetic classification can be increased by sequencing a larger coverage of the 16S rRNA gene or other marker genes inherent within the target microorganism(s).

The use of 16S rRNA amplicon sequencing remains the current method of choice when studying microbial communities. It is currently more cost-effective than whole genome (shotgun) sequencing and provides sufficient coverage and resolution at phylum-class level phylogenetic classification. The use of 16S rRNA sequencing in bioreactor studies have increased significantly in recent times and has become a standard analysis when evaluating complex microbial communities (Aoyagi et al., 2018; Nielsen et al., 2018; Vasquez et al., 2018).

The approach used in any study is dependent on the desired application and outcome. The application of 16S rRNA amplicon sequencing provides a comparative analysis of the microbial community composition and structure across multiple environmental samples and/or at different time points (Jovel et al., 2016). When evaluating microbial community dynamics, it is important not to only quantify the relative abundance of different taxa, but also to track these abundances over time after exposure to different perturbations. This provides an insight into the microbial communities' response to different stimuli and to evaluate the robustness of the biological component. The use of 16S rRNA sequencing thus provides a platform for integrating bioreactor process performance with microbial community dynamics.

3.6.1 Genomic DNA sampling and 16S rRNA amplicon sequencing

Total genomic DNA was extracted from selected samples using NucleoSpin soil genomic DNA extraction kit (Machery-Nagel, Germany) as per manufacturers' instructions. The extracted DNA was checked for purity and concentration determined using a Nanodrop 2000 (Thermo Scientific). Samples were stored at -60°C until further analysis. The extracted genomic DNA was sent to Macrogen (South Korea) for sample preparation, Illumina® MiSeq® sequencing, pre-processing, OTU clustering, and taxonomic assignment (Figure 3.5).

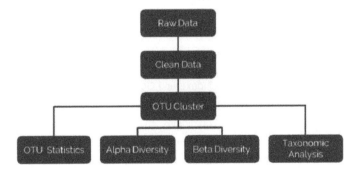

Figure 3.5: Outline of the 16S rRNA sequencing procedure performed by Macrogen on the experimental samples selected for microbial community analysis. Once samples have been sequenced

raw sequencing reads are processed (aligned and filtered). The clean sequencing reads are clustered into unique OTUs and relevant statistical analysis is performed including alpha and beta diversity.

Briefly, dual-index barcoded V3-V4 region sequence libraries were generated by limited cycle polymerase chain reaction (PCR) to yield an approximately 460 bp amplicon, using the bacterial 16S rRNA gene oligonucleotide primers FwOvAd_341F (5'-TCG TCG GCA GCG TCA GAT GTG TAT AAG AGA CAG CCT ACG GGN GGC WGC AG-3') and ReOvAd_785R (5'-GTC TCG TGG GCT CGG AGA TGT GTA TAA GAG ACA GGA CTA CHV GGG TAT CTA ATC C-3'; Dubourg et al., 2016). The degenerate bases (IUPAC code) included in the primer sequences represent more than one nucleotide base possibility and are defined as: N = any base, W = A or T, H = A or C or T and V = A or C or G (Johnson, 2010).

Amplification reactions were performed in a thermal cycler consisting of an initial step denaturation step at 95°C for 3 minutes followed by 25 cycles of denaturation at 95°C for 30 seconds, annealing at 55°C for 30 seconds, elongation at 72°C for 30 seconds, and a final extension step at 72°C for 4 minutes. Amplicon libraries were sequenced on an Illumina® MiSeq® sequencer to yield 300 bp paired-end reads.

3.6.2 Metagenomic OTU picking and taxonomy alignment

The partial 16S rRNA gene sequences was analysed using the bioinformatics pipeline of Quantitative Insights into Microbial Ecology (QIIME) software (www.microbio.me/qiime) as described by Caporaso et al. (2010). Fast length Adjustment of Short reads (FLASH; version 1.2.11) was used to merge paired-end reads (Magoč & Salzberg, 2011). The raw sequence read trimming and filtering as well as operational taxonomic unit (OTU) picking was performed by CD-HIT-OUT at a difference distance cut-off of 0.03 (97% ID similarity at species level) (Li et al., 2012). Lastly, taxonomic assignment of OTUs was performed using the RDP 16S rRNA classifier algorithm, UCLUST (Edgar, 2010).

3.6.3 Metagenomic statistical analysis

The QIIME pipeline also facilitated the determination of the richness indices (Chao 1 estimates), diversity indices (Shannon index), and Goods coverage (Caporaso et al., 2010). Beta diversity analysis was performed using weighted UniFrac algorithm in QIIME and was subsequently analysed by principal coordinates analysis (PCoA) (Lozupone and Knight, 2005). Furthermore, unweighted pair group method with arithmetic mean (UPGMA) clustering based on the weighted UniFrac distance matrix was also performed through QIIME. Phylogenetic evolutionary analysis of OTU sequences was performed using MEGA version 7 software, an integrated tool for conducting manual sequence alignment and phylogenetic tree construction (Kumar et al., 2018).

3.7 Data handling

Reactor performance data and statistical analyses were conducted using Microsoft® excel® 2013. The analytical measurements obtained from the experimental studies were analysed using the following formulae in determining process kinetics and overall process performance. This involved assessing volumetric sulphate reduction rates, volumetric sulphide oxidation rates, substrate utilisation and sulphur recovery.

3.7.1 Kinetic calculations

Sulphate conversion (SC) and lactate conversion (LC):

The sulphate conversion (SC) and lactate conversion (LC) were calculated using the general equation

$$Conversion = \frac{S_0 - S}{S_0} \times 100 \qquad \text{(Equation 3.1)}$$

where S_0 and S represent the feed and residual substrate (sulphate or lactate) concentration (mmol/L), respectively. LC accounts for both oxidation and fermentation of lactate to acetate and propionate.

Expected sulphide (ES):

The expected sulphide was calculated theoretically, based on the amount of sulphate converted to sulphide.

$$ES = \frac{S_0 - S}{3} \qquad \text{(Equation 3.2)}$$

where S_0 and S represents the feed and residual sulphate concentration (mmol/L).

Volumetric sulphate loading rate (VSLR):

This is the product of the feed sulphate concentration and the dilution rate and represents the availability of sulphate for reaction:

$$VSLR = S_0 \times D \qquad \text{(Equation 3.3)}$$

where S_0 is the feed substrate (sulphate or lactate) concentration (mmol/L) and D the dilution rate (1/h).

Volumetric sulphate reduction rates and substrate utilisation and production rates:

The volumetric sulphate reduction (VSRR) and volumetric substrate utilisation rates (mmol/L.h), represented as r_s, were calculated as follows:

$$r_s = (S_0 - S)\, D \qquad \text{(Equation 3.4)}$$

where S_0 and S represent the feed and residual substrate (sulphate, lactate or acetate) concentration (mmol/L) respectively and D is the dilution rate (1/h), the inverse of hydraulic residence time calculated as F/V where F is the feed flow rate and V the reactor volume. VSRR refers to the rate of sulphate removal based on SRB activity while substrate utilisation/production rates describe the utilisation rate of feed substrates (lactate or acetate) and production rate of acetate or propionate in this thesis.

Sulphide conversion:

The conversion of sulphide to elemental sulphur is estimated by the difference between the expected sulphide and that measured:

$$HS^-{}_{removal} = \frac{ES - S}{ES} \times 100 \qquad \text{(Equation 3.5)}$$

where ES and S represents the expected sulphide produced and residual effluent sulphide concentration (mmol/L) respectively.

Biofilm sulphur recovery:

Elemental sulphur recovery through the biofilm can be evaluated based on three parameters.

1) Sulphur recovery based on total sulphate-S load:

$$S^0{}_{recovery} = \frac{S_{FSB}}{SO_{4\,total}} \times 100 \qquad \text{(Equation 3.6)}$$

where S_{FSB} represents the amount of sulphur recovered as elemental sulphur from the biofilm and S_{total} the total amount of sulphate-S load over the duration of the experimental run.

2) Sulphur recovery based on the expected (generated) sulphide-S load:

$$S^0{}_{recovery} = \frac{S_{FSB}}{ES_{total}} \times 100 \qquad \text{(Equation 3.7)}$$

where S_{FSB} represents the amount of sulphur recovered as elemental sulphur from the biofilm and ES_{total} represents the cumulative amount of sulphide-S generated through sulphate reduction over the period operation, between collapsing and harvesting the biofilm.

3) Sulphur recovery based on sulphide removal:

$$S^0{}_{recovery} = \frac{S_{FSB}}{ES_{total} - S_{total}} \times 100 \qquad \text{(Equation 3.8)}$$

where S_{FSB} represents the amount of sulphur recovered as elemental sulphur from the biofilm. ES_{total} and S_{total} represents the cumulative amount of expected sulphide and effluent sulphide over the period of operation between collapsing and harvesting the biofilm.

Chapter 4

Demonstration of the hybrid LFCR

4.1 Introduction

The current chapter describes the development of a novel semi-passive wastewater treatment process for effective remediation of sulphate-rich waste streams. The investigation focuses on the demonstration of the "proof of concept" of a hybrid LFCR process that integrates biological sulphate reduction and partial sulphide oxidation within a single operational unit. Furthermore, it is hypothesised that the inclusion of carbon microfibers, within a hybrid LFCR, facilitate the attachment and retention of a sulphate reducing microbial community within the bulk volume of the reactor, while the hydrodynamic properties present in the LFCR facilitate the establishment of a discrete anaerobic zone supporting biological sulphate reduction by SRB and a microaerobic zone in which the biofilm creates a suitable microenvironment for partial sulphide oxidation by sulphur oxidising bacteria (SOB).

The intention of this study was to assess the performance in terms of sulphate reduction efficiency, sulphide oxidation efficiency and the potential for sulphur recovery as a value added product as well as for water remediation. Through this, the feasibility of the process to be applied at larger scale for the treatment of sulphate-rich waste streams as part of a semi-passive bioprocess is investigated.

The specific objectives addressed in this chapter were as follows:

1. Determine hydrodynamic regime within the 2 L LFCR configuration and whether it conformed to the conceptual model previously described by Mooruth (2013).
2. Demonstrate the "proof of concept", integrating simultaneous biological sulphate reduction and partial sulphide oxidation within a single LFCR configuration.
3. Assess the harvesting system for recovery of the floating sulphur biofilm, that minimises the time required for the biofilm to reform.

4. Evaluate overall process kinetics for achieving high sulphate reduction and partial sulphide oxidation through a floating sulphur biofilm for elemental sulphur recovery.
5. Assess the effect of hydraulic residence time on process kinetics in terms of volumetric sulphate reduction and partial sulphide oxidation as well as sulphur recovery.
6. Investigate the impact of periodic biofilm disruption on process performance.

This chapter is arranged as follows: the demonstration of the hybrid LFCR is discussed, followed by the effect of hydraulic residence time and biofilm disruption on process performance. The respective experimental methods, results and discussions pertaining to each experiment are then presented.

4.2. Hydrodynamics of the LFCR

The hydrodynamic performance of a biological reactor is an important design characteristic since it directly affects the process efficiency (Khalekuzzaman et al., 2018). The lack of a fundamental understanding of reactor hydrodynamics can lead to process complications (Mooruth, 2013). This was the case in the original LFCR design, where the hydrodynamics of the system were not well understood, resulting in poor reactor performance (Pulles & Howard, 2009). Mooruth et al. (2013) performed an extensive hydrodynamic analysis on a 25 L LFCR configuration. The study resulted in the optimisation and re-design of the LFCR to facilitate optimal conditions for partial sulphide oxidation, highlighting key features of the mixing regime within the LFCR. In addition, Mooruth et al. (2013) demonstrated the feasibility of achieving high partial sulphide oxidation via the formation of a FSB as well as the key role of fluid dynamics on reactor performance. The study formed part of the foundation that led to the conceptualisation of the hybrid bioprocess. Thus, it was important to confirm that the fluid dynamics present in the 2 L LFCR, used in the current study, was consistent with the 25 L configuration previously described by Mooruth (2013).

To evaluate the fluid mixing profile in the LFCR, a phenolphthalein dye tracer study described in Section 3.4 was conducted (Figure 4.1). The tracer study showed that the mixing in the LFCR was governed by the feed velocity (advective transport) at the reactor inlet, which caused some brief, localised turbulent eddies. The absence of turbulent mixing and a slight density difference caused the acid feed to sink to the bottom of the channel. A dead zone was observed at the front corner of the reactor and the acid front moved along the floor of the reactor with a laminar parabolic profile (Figure 4.1 B and C). After 30 minutes (Figure 4.1 B), the acid front reached the back wall of the reactor, resulting in the vertical displacement of the HCl layer. As time progressed, convective transport (a combination of advective and diffusive transport) became predominant (Figure 4.1 D and E), from the front and back of the reactor towards the middle, until the entire bulk volume turned colourless (macromixing time).

Figure 4.1: Photographic images of the tracer study showing the progression of mixing over time in the 2 L LFCR configuration at a 2 day HRT at ambient temperature. A) 0 min B) 30 min C) 48 min D) 114 min E) 135 min F) 145 min. Direction of flow proceeding from left to right. The internal length dimension (250 mm) is shown in image A for scale. Features of the LFCR such as the sampling ports and heating element can be observed.

The study concluded that the fluid dynamics in the LFCR is primarily governed by passive mixing through a combination of advective and diffusive transport. The relatively slow linear velocity and absence of turbulent mixing meant there was minimal disturbance at the surface of the reactor. Mooruth (2013) also indicated very low levels of turbulence existed within the LFCR. This was critical to ensure suitable conditions that favoured the development of a floating sulphur biofilm at the air-liquid interface. Additionally, the results revealed that the LFCR achieved complete mixing times that were considerably shorter (2h 25 min) than the hydraulic residence time (2 day) tested. These findings are consistent with the conceptual fluid dynamic model previously described by Mooruth (2013), shown in Figure 4.2.

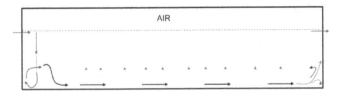

Figure 4.2: Conceptual model of the fluid dynamics regime present within the 25 L LFCR, adapted from Mooruth, (2013). The sinking influent and effluent (green), entrance dead zone (red arrows), laminar parabolic flow (black arrows) and back corner dead zone (orange, maroon, blue arrows) are displayed. The vertical pink arrows represent the y-directional fluid movement due to diffusive transport.

Ntobela & Chibwana, (2016) conducted a saline tracer study in the LFCR which revealed that it exhibited a similar residence time distribution (RTD) profile to that of a CSTR, which further supported the findings that the LFCR can be considered as a relatively well-mixed system, but with low surface renewal. Furthermore, the observed fluid dynamics demonstrated the suitability of the reactor design for the desired application as it facilitates complete mixing within the bulk volume of the reactor with limited turbulence at the surface, promoting ideal conditions for sulphate reduction and the formation of the floating sulphur biofilm at the air-liquid interface. In addition, the fluid mixing regime was preserved within 2 L LFCR and performed similarly to that described by Mooruth, (2013).

4.3 Demonstration of the hybrid LFCR reactor

4.3.1 Experimental approach

For the demonstration of the proof of concept, the following experimental procedure was used. A 2 L LFCR was inoculated with a mixture of active SRB and SOB cultures (Section 3.1.2) and fed by a speed controlled peristaltic pump with modified Postgate B medium containing 1 g/L sulphate at a 4 day HRT (dilution rate: 0.104/h). Lactate served as the electron donor and carbon source and was supplemented at 11 mM to achieve a COD:sulphate ratio of 0.7. The temperature was controlled at 28°C by the internal heating coil connected to a circulatory heating bath. As described in Section 3.1.3, samples (2 mL) were taken daily from the middle and lower sample ports in the first and third columns (FM, FB, BM and BB), as well as from the effluent port. The pH and oxidation reduction potential (ORP) as well as the sulphide concentration were measured immediately, as described in Section 3.2.2 and 3.2.3. The remainder of the sample was prepared for VFA and anion analysis by chromatography (Section 3.2.4 and 3.2.6). FSB formation was visually observed and once a thick, stable biofilm had been formed it was periodically disrupted and harvested. The FSB was disrupted by physically disrupting the biofilm and allowing the fragments to settle on the harvesting screen. The sulphur product was recovered by removing the harvesting screen and collecting the accumulated biofilm. Thereafter, the biofilm was dried at 37°C for 48 hours, weighed and stored for further elemental analysis (Section 3.3).

4.3.2 Results and discussion

4.3.2.1 Initial start-up and performance of the hybrid LFCR

The LFCR was inoculated with an active SRB mixed microbial culture with an initial sulphide concentration of approximately 7 mmol/L (230 mg/L). The sulphide concentration decreased rapidly over the first 24 h as a result of unimpeded oxygen mass transfer across the liquid surface, resulting in sulphide oxidation. Within 24 hours, a thin, but complete biofilm was observed covering the entire surface. The colonised reactor with the well-developed FSB at the air-liquid interface is shown in Figure 4.3. Once the biofilm had formed, the dissolved sulphide concentration in the bulk liquid began to increase steadily (Figure 4.4), from around 0.5 mmol/L (16 mg/L) to 4.6 mmol/L (152 mg/L) by day 34. This was an indication of SRB activity, which was further confirmed by the concomitant decrease in residual sulphate concentration from 10.86 mmol/L (1042 mg/L) within the feed to 3.89 mmol/L (373 mg/l) in the effluent (Figure 4.4 B) by day 34, corresponding to a sulphate conversion of approximately 64%.

Figure 4.3: Demonstration of the hybrid LFCR showing the set-up of a colonised reactor with a well-developed floating sulphur biofilm completely covering the exposed liquid surface after 4 days after disruption of the FSB.

On day 34 a controlled biofilm disruption and harvest resulted in the rapid decrease in sulphide concentration and slight increase in residual sulphate concentration. As the FSB regenerated at the air-liquid interface, the sulphide concentration increased to 3.2 mmol/L by day 45 after which the biofilm was disrupted again. Mooruth (2013) hypothesised that as the biofilm develops and matures it acts as a barrier, impeding oxygen mass transfer across the air-liquid interface. The colonisation of the carbon fibres resulted in the increased retention of biomass and by day 60 the residual sulphate concentration (Figure 4.4) decreased to 1.0 mmol/L (<100 mg/L) reaching a VSRR and sulphate conversion of 0.11 mmol/L.h (10.56 mg/L.h) and 96%, respectively.

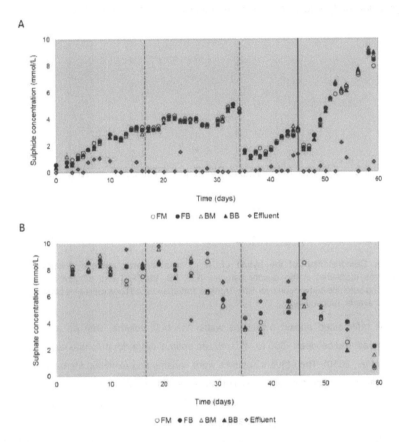

Figure 4.4: Start-up and demonstration of the hybrid LFCR showing A) residual sulphate and B) dissolved sulphide concentration profiles measured through the reactor volume (FM, FB, BM, BB) and effluent over time. Biofilm disruption and harvest events are indicated by vertical dotted and solid lines, respectively. The reactor samples represent the front-middle (FM), front-bottom (FB), back-middle (BM) and back-bottom (BB) sampling ports.

The mean residual sulphate concentration (Figure 4.4 A) measured in the reactor samples (FM, FB, BM, BB) and effluent was consistent throughout the study with minimal variation, indicating limited complete oxidation of sulphide to sulphate occurred. Partial sulphide oxidation (Figure 4.4 B) was efficient, with low concentrations measured in the effluent. The average sulphide conversion, between day 30 and 60, was over 90%. This was calculated based on the expected amount of sulphide generated based on the residual sulphate converted and the final effluent sulphide concentration. In addition, thiosulphate concentrations remained below detection limits for the duration of the experiment. Collectively, these findings suggest that partial sulphide oxidation to elemental sulphur was favoured throughout the study, with limited complete oxidation of sulphide to sulphate.

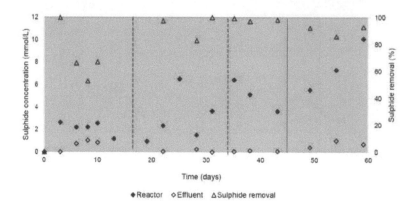

Figure 4.5: Demonstration of the hybrid LFCR process showing average dissolved sulphide concentrations measured in the reactor samples (FM. FB, BM, BB) and effluent as well as the corresponding sulphide conversion over time. Vertical dotted and solid lines represent biofilm disruption and harvest events, respectively.

The LFCR maintained anoxic conditions within the bulk volume, with an average redox potential measured between -350 to -410 mV, an optimal range for sulphate reducing activity (Sheoran et al., 2010). The effluent samples were variable and exhibited increased oxidation of the oxidation reduction potential (ORP) measurements throughout the duration of the study. This was expected since the effluent is exposed to the aerobic zone at the surface as it flows through the exit port of the reactor. The ability of the LFCR to maintain anoxic conditions within the bulk volume, despite the surface being open to the atmosphere, was critical to achieve simultaneous sulphate reduction and partial sulphide oxidation.

As shown in Figure 4.6, the pH increased initially in the bulk volume of the reactor (pH 7-7.5) with an additional increase observed in the effluent (pH 7.5-8). The initial pH increase was attributed to SRB activity as a result of alkalinity (bicarbonate) production (Reaction 2.9) while the additional pH increase observed in the effluent was attributed to partial sulphide oxidation, where hydroxyl ions are released as a by-product. Since the experiments were conducted using a feed at neutral pH, these results confirm the generation of alkalinity and highlights the potential of the system to neutralise wastewater streams.

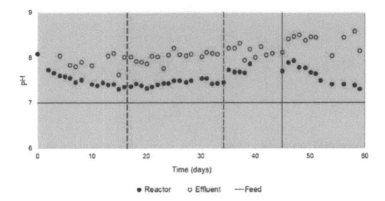

Figure 4.6: Average pH data measured in reactor samples (FM, FB, BM, BB) and effluent over the initial start-up and demonstration phase. Minimal variation (≤ 0.05) across reactor samples, vertical dotted and solid lines represent biofilm disruption and harvest events, respectively.

4.3.2.2 Lactate metabolism in the hybrid LFCR

The study assessed the anaerobic metabolism of lactate within the hybrid LFCR via monitoring VFA concentration profiles. Lactate, can be metabolised via complete oxidation (Reaction 4.1), incomplete oxidation (Reaction 4.2) or through fermentation (Reaction 4.3) by a wide range of microorganisms, (Bertolino et al., 2012; Oyekola et al., 2012):

$$2\ lactate + 3\ SO_4^{2-} \rightarrow 6\ HCO_3^- + 3\ HS^- + H^+ \quad \text{(Reaction 4.1)}$$

$$2\ lactate + SO_4^{2-} \rightarrow 2\ acetate + 2\ HCO_3^- + HS^- + H^+ \quad \text{(Reaction 4.2)}$$

$$3\ lactate \rightarrow acetate + 2\ propionate + HCO_3^- + H^+ \quad \text{(Reaction 4.3)}$$

As shown in Figure 4.7, throughout the experiment, the residual lactate concentration was below the detection limit. Acetate accumulated as the main by-product, while propionate concentrations were consistently low. This indicated that incomplete lactate oxidation (Reaction 4.2) was the dominant metabolic pathway. Furthermore, the low propionate concentrations suggested very little fermentation (Reaction 4.3) occurred throughout this study.

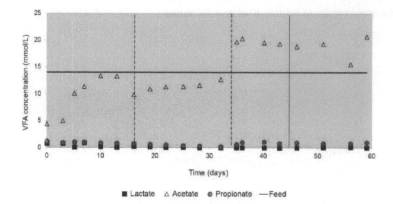

Figure 4.7: Average VFA data during initial start-up and demonstration showing the measured residual lactate, acetate and propionate concentrations and the lactate feed concentration, vertical dotted and solid lines represent biofilm disruption and harvest events, respectively.

These results were consistent with previous studies that investigated the VFA concentration profile in sulphate reducing reactors fed with lactate as a sole carbon source (Oyekola et al., 2012; Bertolino et al., 2012). Measured acetate concentrations exceeded the theoretical values based on reaction stoichiometry (amount of lactate supplied in the feed) from day 30 (Figure 4.7). This was attributed to the metabolism of yeast extract and citrate within the feed. Citrate metabolism is described in literature, with acetate as a possible reaction product (Stams et al., 2009). Yeast extract is traditionally supplemented into microbial growth media as a source of nitrogen, vitamins and trace metals. However, it also contains carbohydrates (4-13%) which can be broken down to acetate (EURASYP, 2015). The accumulation of acetate exceeding the theoretical amount expected based on feed lactate concentration indicates that it is unlikely that complete lactate oxidation (Reaction 4.1) took place.

The hybrid LFCR process achieved a final sulphate conversion of 96% with a corresponding VSRR of 0.11 mmol/L.h by day 60, when operated at a 4 day HRT (Figure 4.4). These results are consistent with data obtained using the same microbial community in conventional CSTRs under similar operating conditions (Oyekola et al., 2012). Between day 35 and 60 complete sulphide conversion (95-100%) was achieved with the recovery of 30% of the added sulphur as elemental sulphur by harvesting the biofilm. A detailed study by Mooruth (2013) evaluated the chemical reactions that determine sulphur speciation in the LFCR and its dependence on pH and colloidal sulphur concentration. The study concluded that colloidal sulphur concentrations >2 mM in the pH range of 8.1 and 9.5 would result in polysulphide formation. Since the current system operated below pH 8 it is more likely that the fraction of elemental sulphur that was not recovered through the biofilm was suspended in solution. This was confirmed by HPLC analysis, which detected colloidal sulphur present in the effluent, as well as the accumulation of sulphur particles and biofilm fragments which settled in the effluent pipe and reservoir. The accumulation of sulphur released in the overflow could be recovered by gravity sedimentation, while any formation of polysulphides downstream can be rapidly converted to sulphur by decreasing the pH. This is a standard procedure applied in conventional sulphur recovery treatments (i.e. Thiopaq® process) (Cai et al., 2017).

4.3.2.3 Floating sulphur biofilm operation

An important component to the successful operation of the hybrid LFCR, was the management of the FSB to ensure optimal partial sulphide oxidation and sulphur recovery. The development of the floating sulphur biofilm at the air/liquid interface over time is illustrated in Figure 4.8. After 24 hours (Figure 4.8 B) a thin layer of biofilm covered the surface, which continued to develop as the biomass and elemental sulphur content increased. By day 7 (Figure 4.8 D) the biofilm had matured becoming completed oxygen limiting resulting in the decrease in sulphide oxidation activity. This was consistent with the high accumulation of sulphide within the reactor. The biofilm was disrupted on day 7, with the broken fragments of biofilm allowed to settle onto the harvesting screen, situated just below the air-liquid interface (Figure 4.8 D). The process was then repeated and after 24 h the biofilm had reformed at the surface.

The recovery of the biofilm was an essential feature of the hybrid LFCR process. Previous studies by Mooruth, (2013) employed a harvesting strategy in which the biofilm was disrupted and allowed to settle at the bottom of the reactor. Thereafter, the bulk reactor volume would be drained, and biofilm recovered. This strategy was impractical, particularly for larger scale application, due to the periodic down time of draining the reactor and having to restart the start the process (Mooruth, 2013). In the current study, the addition of a harvesting screen was a critical component to the success of operating the hybrid LFCR process, serving as an effective mechanism for the recovery of the FSB and more importantly, minimal disturbance to the anaerobic zone. It also permitted the biofilm to be disrupted intermittently, facilitating multiple cycles before a harvest was required.

Throughout the study three distinctive stages of biofilm consistency was observed as the biofilm matured over time. This was previously described by Molwantwa, (2008) as the thin, sticky and brittle phases. These stages play an important role in the development and functioning of the FSB for the desired application. A descriptive model accounting for the structure and function of its development is best described by Mooruth, (2013) as follows: Heterotrophic aerobic and micro-aerophilic bacteria establish and lay down an organic carbon matrix (EPS) at the air-liquid interface, which forms the framework of the biofilm. As the biofilm matures a steep DO and redox gradient at the surface develops, where the biofilm functions as a barrier to oxygen mass transfer. This establishes a microaerobic zone within the biofilm, where pH and redox conditions favour the partial oxidation of sulphide, which is present within the bulk volume and continuously migrates towards the FSB. The constructed EPS layer serves as an attachment site for the colonisation of SOB resulting in the increased deposit of elemental sulphur in the biofilm, visualised by the distinctive yellow/white coloration. As the FSB continues to mature over time, it becomes oxygen limiting at which the rate of sulphide oxidation is reduced. Thus, a biofilm disruption is required periodically during operation to re-establish sulphide oxidation conditions and to ensure maximum sulphur recovery.

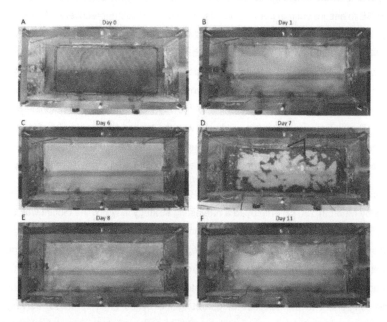

Figure 4.8: Photographic recording of the floating sulphur biofilm formation at the air-liquid interface over time, on day 7 (D) a biofilm disruption occurred by physically disrupting the biofilm and allowing the fragments to settle on the mesh screen positions just below the interface.

4.3.2.4 Assessing the FSB and biomass attachment on carbon microfibers

The hybrid LFCR facilitated the partitioning of two reactive zones, namely; the anaerobic zone within the bulk volume and the aerobic zone at the air-liquid interface. These zones promoted the development of distinctive microbial communities that are critical for optimal process operation, including the colonisation of the carbon fibres as a support matrix for biomass attachment of SRB and the formation of the FSB. In this study the FSB as well as the attachment of biomass onto carbon fibres were visualised through scanning electron microscopy (SEM), as described in Section 3.5, and are presented in Figure 4.9.

Previous studies by Molwantwa, (2008) and Mooruth, (2013) performed comprehensive structure/function analysis on the FSB using a combination of physicochemical and molecular techniques. These studies described the FSB as an intricate network housing distinct physiological compartments with microbial population and spatial structural differentiation. The complexity of the FSB was defined by the incorporation of a diverse microbial community encapsulated within an EPS matrix containing pores and channels, typical characteristics associated with biofilms.

Results from the current study revealed a diverse microbial community populated the FSB by the presence of diverse bacterial morphologies observed in Figure 4.9 A. Sulphur deposits were also visualised on the outer membranes of the bacterial structures, a strong indication of SOB present within the FSB (Figure 4.9 A). The presence of sulphur globules excreted on the outer membranes of SOB species has been well documented (Cai et al., 2017). The mechanisms involved in the formation and consumption of sulphur globules in photosynthetic sulphur bacteria (PSB) and green sulphur bacteria (GSB) are largely unknown. It is formed as an intermediate during the incomplete oxidation of sulphide under oxygen limiting conditions as a storage mechanism. When reducing equivalents are required (sulphide is depleted) and sufficient amounts of an electron acceptor is available (oxygen or nitrate), the sulphur globules are oxidised completely to sulphate (Holkenbrink et al., 2011). SEM-EDS analysis (Figure 4.9 B) of the FSB confirmed partial sulphide oxidation, with the presence of highly concentrated elemental sulphur (100%) deposits detected across selected points (Figure 4.9 B; S1-S3).

Figure 4.9 D and C confirmed biomass attachment and colonisation of the carbon microfibers. The attached biomass on the carbon fibres were distinctive from that associated with the FSB with cell morphologies resembling rod and vibrio forms, a characteristic feature of many SRB species. These results together with the sulphate reduction performance support the study by van Hille et al. (2015), which demonstrated enhanced VSRR achieved using carbon fibres as an effective internal support matrix for biomass attachment.

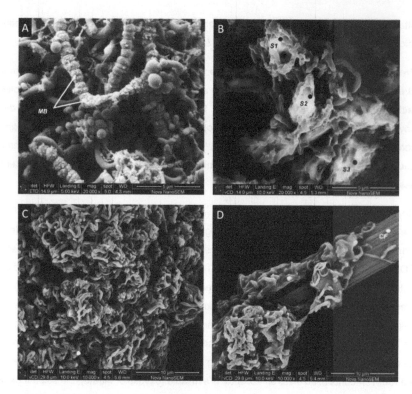

Figure 4.9: Scanning electron microscopy images of the A) FSB microbial community showing bacterial morphologies with membrane bound sulphur deposits (MB) B) SEM-EDS imaging and elemental composition of the floating sulphur biofilm; EDX point analysis of S1, S2 and S3 detected 100% sulphur (all elements were normalised), C) biomass associated with the carbon fibres showing EPS encapsulated bacterial cells, and D) microbial cells and biofilm attached to the surface of carbon fibre (CF) as a support matrices. Diverse cell morphologies, including cocci-, bacillus- and vibrio-shaped cells, are observed clustered in an intricate network within the FSB and encapsulated in EPS in the biofilm attached to the CF.

4.4 Evaluating the effect of hydraulic residence time

Following the successful start-up and demonstration of the hybrid LFCR process, a study was initiated to further evaluate system performance. An important operational parameter of BSR technologies, after the selection of a cost-effective electron donor, is an optimal hydraulic residence time (HRT) and the resilience of the system to its fluctuation. This is particularly important when evaluating the feasibility of a process for effective treatment since it influences the hydraulic conditions in the reactor and the contact time of the contaminant with the active microbial community (Neculita et al. 2008; Vasquez et al., 2018). The criteria by which BSR treatment is considered a feasible option typically depends on economics as well as the ability of the process to adequately remove sufficient sulphate in a given time such that the discharge of the treated water meets the desired regulatory requirements (Bowell, 2004). Previous studies have reported the effect of HRT in BSR systems with a primary focus on evaluating sulphate and metal removal. In this study the effect of HRT on the performance of the hybrid LFCR was performed.

4.4.1 Experimental approach

A range of HRTs (3, 2, 1 and 0.5 day) were tested by changing the feed dilution rate into the reactor. Prior to commencement of the study, the flow rate was changed to a 3 day HRT and the FSB was harvested. After approximately 3 residence times (RTs) at 3 day HRT, equivalent to 9 days of continuous operation, the biofilm was disrupted and the system was allowed to proceed for an additional 3 RTs. A second biofilm disruption followed and the accumulated biofilm over the duration of 6 RTs (18 days) in total was harvested. Upon harvesting the biofilm, the feed rate was increased to achieve the next HRT to be tested and the process was repeated. It was important that the total volume treated (sulphate load) was kept constant across all HRTs evaluated. Therefore, a total of 6 RTs (6 reactor volumes) of system operation was kept constant for each HRT tested. This was to ensure consistency across the study, particularly when assessing the performance of the FSB for sulphur recovery.

4.4.2 Results and discussion

The experimental data collected over the duration of the study is shown in Figure 4.10 - Figure 4.18. A consistent trend was observed when the floating sulphur biofilm (FSB) was disrupted and harvested, the sulphide concentration decreased rapidly over 24 hours before gradually increasing as the FSB re-formed at the surface. The highest sulphide concentration was measured at 8.2 mmol/L when operated at a 3 day HRT. As the HRT was incrementally decreased, a decline in the maximum sulphide concentration was observed, reaching 3.7 mmol/L at a 12 h HRT.

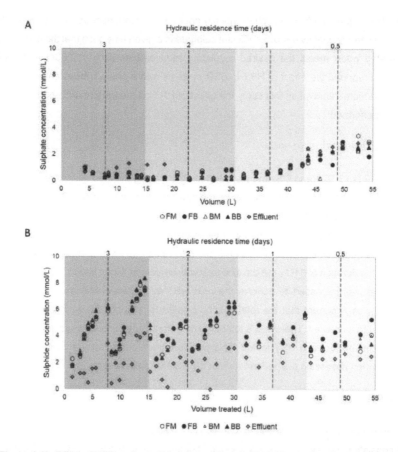

Figure 4.10: Effect of HRT on the performance of the hybrid LFCR showing A) residual sulphate and B) dissolved sulphide concentration profiles as a function of volume treated. Biofilm disruption and harvest events are indicated by vertical dotted and solid lines, respectively. A change in HRT is indicated by the transition in shading intensity and was accompanied by a biofilm harvest.

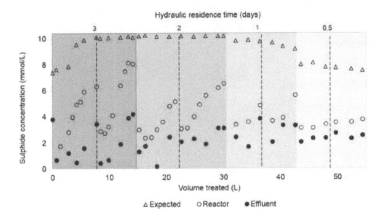

Figure 4.11: Expected sulphide generated in the absence of its partial oxidation to elemental sulphur, based on sulphate reduction and mean dissolved sulphide concentration measured in the reactor and effluent. Biofilm disruption and harvest events are indicated by vertical dotted and solid lines, respectively. A change in HRT is indicated by the transition in shading intensity and was accompanied by a biofilm harvest.

Figure 4.11 shows that the reactor sulphide concentration tends towards the expected sulphide concentration. The residual sulphate concentration was consistently low (± 0.2 mmol/L) for both 3 and 2 day HRT which corresponded to a sulphate conversion of approximately 98%. A gradual increase in sulphate concentration (1.2 mmol/L) was observed at a 1 day HRT, resulting in a decreased sulphate conversion of 88%. The most noteworthy increase in sulphate concentration was observed at a 0.5 day HRT, with a residual sulphate concentration of 2.8 mmol/L and corresponding sulphate conversion of 73%.

The accumulation of aqueous sulphide is best represented as a function of time as seen in Figure 4.12. The longer operation period before disruption at a 3 day HRT (9 days) compared to that at a 12 h HRT (1.5 days) allowed for higher sulphide accumulation. This was consistent with the formation and longer presence of the biofilm at the surface over the period. As the biofilm forms it acts as a barrier at the air-liquid interface, impeding oxygen mass transfer into the bulk volume (Mooruth, 2013). When the biofilm was disrupted or harvested, the barrier was disrupted and an increase in unimpeded oxygen penetration into the bulk volume resulted in the rapid oxidation of the available aqueous sulphide. After 24 hours following a biofilm disruption or harvest event, a thin but complete biofilm layer covered the entire surface of the reactor. Over the next 3 HRTs, the biofilm increased in thickness and sulphur content followed by an increase in aqueous sulphide concentration in the bulk volume. FSB disruption and harvesting had minimal impact on the residual sulphate concentration, in comparison to the dissolved sulphide concentration, which indicated that the sulphate reduction activity was not affected. In addition, there was minimal re-oxidation of the sulphide to sulphate. These findings were consistent with the results obtained in Section 4.3 during the initial start-up and demonstration phase.

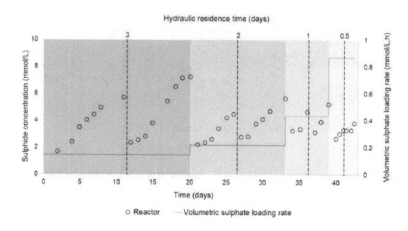

Figure 4.12: Average dissolved sulphide concentration profile in the hybrid LFCR reactor samples (FM, FB, BM, BB) as a function of time. Biofilm disruption and harvest events are indicated by vertical dotted and solid lines, respectively. A change in HRT is indicated by the transition in shading intensity and was accompanied by a biofilm harvest.

The sulphide concentration was notably higher at the end of each HRT experimental run, compared to the concentration reached after the mid-way biofilm disruption (Figure 4.12). This was a result of biofilm remnants that remained at the interface after disrupting the biofilm (dotted vertical line), but not after harvesting (solid vertical line). The presence of these fragments facilitated a more rapid regeneration of the biofilm which enabled faster accumulation of aqueous sulphide, possibly by providing an attachment site or maintaining a high concentration of microbial cells that are responsible for the development of the FSB. These explanations coincided with the faster regeneration of the biofilm and higher sulphide concentration measured after a biofilm disruption event (dotted vertical line line) in comparison to a biofilm harvest (solid vertical line).

The operating pH of a BSR process does not only affect the SRB activity but also determines sulphide speciation. Sulphide in its gaseous form (H_2S) is highly toxic and can have a detrimental effect on system performance which may lead to process failure (Oyekola et al., 2012; Sánchez-Andrea et al., 2014). In this study, the feed was kept at a pH of 7 to ensure sulphide would predominantly remain in the aqueous phase (HS^-) and to provide suitable conditions that favour both sulphate reduction and partial sulphide oxidation. In addition, this ensured that sulphide would predominantly exist in solution with minimal loss to the surrounding atmosphere. The average pH data measured in the reactor samples (FM, FB, BM, BB) remained relatively stable between pH 7 and 7.5 for the duration of the experiment. There was an initial increase in pH within the reactor volume with an additional increase observed in the final effluent. This initial increase was attributed to the bicarbonate generated as a consequence of biological sulphate reduction, which acted as a buffer in the system. While the increase in pH observed in the final effluent was a result of hydroxyl ions released as a consequence of partial sulphide oxidation. The decrease in HRT resulted in a downward trend in pH as the HRT was reduced to 0.5 day. This corresponded well with the observed decrease in sulphate conversion, shown in Figure 4.15, and consequent reduction in bicarbonate generation.

The ORP measured in the effluent remained relatively constant throughout the study ranging between -380 and -410 mV for the duration of all HRTs tested. Though the reactor volume was exposed to the surrounding atmosphere, the ability to establish and maintain appropriate anoxic (reducing) conditions suitable for sulphate reduction was an important requirement for the successful operation of the hybrid LFCR.

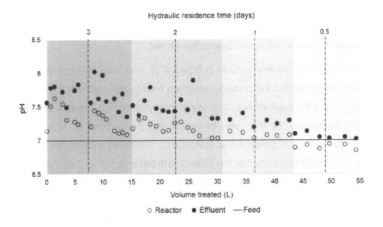

Figure 4.13: Mean pH data measured in the reactor and effluent samples. Biofilm disruption and harvest events are indicated by vertical dotted and solid lines, respectively. A change in HRT is indicated by the transition in shading intensity and was accompanied by a biofilm harvest.

VFA concentrations (Figure 4.14) were measured over the duration of the study to assess the impact of HRT on the metabolic pathways present in the hybrid LFCR. The depletion of lactate and concomitant accumulation of acetate strongly suggested incomplete lactate oxidation was the primary metabolic pathway at a 3 and 2 day HRT. This was consistent with the low sulphate concentration observed in Figure 4.10 A. An increase in sulphate concentration was observed at a 1 and 0.5 day HRT, which corresponded with the simultaneous increase in propionate concentration and gradual decline in acetate concentration, a strong indication of lactate fermentation (Bertolino et al., 2012)

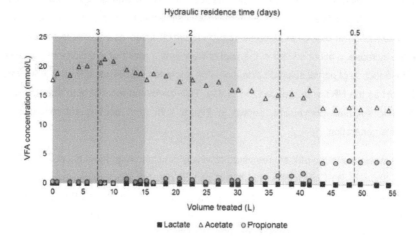

Figure 4.14: Volatile fatty acid concentration profile showing the measured lactate, acetate and propionate concentrations, respectively, within the LFCR. Biofilm disruption and harvest events are indicated by vertical dotted and solid lines, respectively. A change in HRT is indicated by the transition in shading intensity and was accompanied by a biofilm harvest.

The reduction in sulphate conversion seen in Figure 4.15 was a result of operating the LFCR at a shorter HRT which limited the contact time of the feed (substrate) with the active culture, affecting the activity of the sulphate reducing community. Additionally, the increase in available electron donor (increased loading rate) promoted the proliferation of opportunistic, fast growing fermentative microorganisms that compete for lactate as a carbon source at higher dilution rates. This result suggests that the loading rate became higher than that at which the SRB community can adequately assimilate all the carbon for sulphate reduction favouring the growth of fast growing organism. This resulted in a shift in the dominant metabolic pathway of lactate utilisation in the reactor, where the increased fermentation of lactate, resulted in a decrease in lactate oxidation for sulphate reduction.

The overall system performance at each HRT is summarised in Table 4.1, while steady state kinetics of both sulphate reduction and sulphide oxidation are shown in Figure 4.15. The volumetric sulphate reduction rate (VSRR) increased linearly from 0.14 to 0.63 mmol/L.h with an increase in the volumetric sulphate loading rate (VSLR). Sulphate conversion was most efficient when operated at a 3 and 2 day HRT where 98% conversion was achieved. However, this declined when the HRT was reduced to a 1 day and 12 h HRT. Despite the reduction in conversion, the system maintained high conversion of 80 and 73% and was able to sustain a linear increase in VSRR at a 1 day and 12 h HRT, respectively. The fact that the system could achieve steady state even at a 12 h HRT and did not experience any cell wash out or system failure, demonstrated the robustness of the process to HRT fluctuation. The high biomass retention achieved by attachment to carbon microfibers enabled high VSRRs to be achieved at short HRTs. Operating a continuous stirred-tank reactor (CSTR), without biomass retention or partial sulphide oxidation, at an HRT below 1 day resulted in system failure as a consequence of SRB washout and proliferation of fermentative microorganisms (Oyekola et al., 2009).

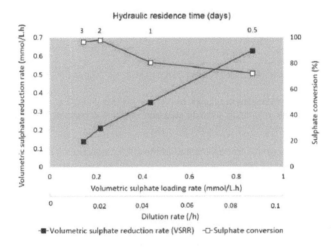

Figure 4.15: Steady state kinetics of sulphate reduction in the hybrid LFCR as a function of HRT. Kinetic data presented are the volumetric sulphate reduction rates and sulphate conversion.

The partial sulphide oxidation in the hybrid LFCR was governed by the available sulphide concentration within the bulk volume, where the generated sulphide product of sulphate reduction serves as the substrate toward elemental sulphur production. Therefore, the performance of the FSB was assessed based on average volumetric sulphide oxidation (VSOR) shown in Figure 4.16. This was calculated based on the expected amount of sulphide generated through sulphate conversion and the final sulphide concentration measured in the effluent over the duration of each HRT.

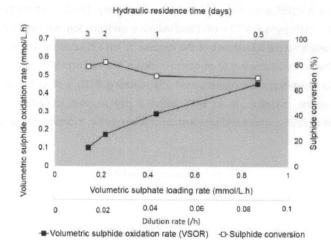

Figure 4.16: Steady state kinetics of sulphide oxidation in the hybrid LFCR as a function of HRT. Kinetic data presented are the volumetric sulphide oxidation rate and corresponding sulphide conversion efficiency.

The results followed a similar trend exhibited by the sulphate reduction performance. However, the increase in VSOR was less pronounced at a 1 day and 12 h HRT. This directly coincided with the observed decrease in sulphate reduction at the shorter HRT range. These results highlight the dependency of sulphide oxidation on the performance of sulphate reduction occurring in the hybrid LFCR system. Sulphide conversion was still effective at the shorter HRTs, achieving 71 and 70% at a 1 day and 12 h HRT respectively, compared to 78-82% at the 2 and 3 day HRT.

Table 4.1: Summary of the effect of hydraulic residence time on steady state kinetics in the hybrid LFCR showing overall sulphate reduction and sulphide oxidation performance.

HRT (days)	Volumetric sulphate loading rate (mmol/L.h)	Volumetric sulphate reduction rate (mmol/L.h)	Sulphate conversion (%)	Volumetric sulphide oxidation rate (mmol/L.h) [a]	Sulphide oxidation (%) [a]	FSB sulphur recovery (%)	Gap-S (%) [b]
2 L lactate-fed							
3	0.14	0.14	97	0.10	78	24	51
2	0.22	0.21	98	0.17	82	27	52
1	0.43	0.35	81	0.29	71	23	21
0.5	0.87	0.63	73	0.45	70	16	15

[a] Average VSOR and sulphide conversion based on the expected sulphide generated (sulphate conversion) and final effluent concentration.
[b] Sulphur predominantly in the form of colloidal sulphur and biofilm fragments in the effluent

Sulphide oxidation for sulphur recovery has been applied in numerous reactor configurations under precise operational control (Syed et al., 2006; Sun et al., 2017). These systems are typically active processes carried out in liquid suspension and require downstream sulphur separation using a clarifier or sedimentation tank (Cai et al., 2017). Alternatively, centrifugation has been applied in the Thiopaq™ process, at industrial scale (Janssen et al., 2000). However, this is expensive, with high energy and maintenance requirements and is not a viable approach to ARD waste streams, particularly in remote areas. One of the most attractive aspects of the hybrid LFCR is the recovery of elemental sulphur as part of a FSB, as it is more beneficial to harvest the sulphur than for it to remain as colloidal sulphur. The mesh-screen situated just below the FSB proved to be an effective method for harvesting the biofilm. It allowed for biofilm disruption and accumulation before a biofilm harvest was required. The harvesting of the biofilm had minimal impact on sulphate reduction and allowed the system to operate continuously without disturbance to process performance.

Elemental composition analysis of the biofilm harvested at the end of each HRT is shown in Figure 4.17. The total sulphur load (4.2 g) fed into the system for each HRT tested was kept constant to determine its effect on the performance of sulphur recovery.

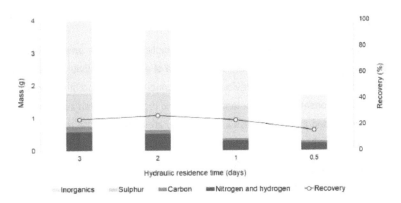

Figure 4.17: Effect of HRT on floating sulphur biofilm recovery over 6 RTs, showing the composition of inorganics, sulphur, carbon, nitrogen and hydrogen in the amount of biofilm harvested and sulphur recovery.

There was a decreasing trend in the amount of biofilm harvested as HRT decreased. A total of 4 g (dry mass) was harvested at a 3 day HRT (18 days operation), compared to 1.7 g at a 12 h HRT (3 days operation). The sulphur recovery from the FSB remained relatively constant as the HRT was reduced from 3 to 1 day HRT (24 to 23%). However, the recovery declined to 16% at a 12 h HRT.

The sulphur content (% composition) of the biofilm increased from 25% to 40% as the HRT was reduced from a 3 to 1 day HRT, owing to a reduction in the carbon and inorganic fractions accumulated with decreasing HRT. Further reduction to 12 hours (39% S) had no substantial impact on FSB composition. The carbon content was considerably low compared to the sulphur fraction and decreased from 4.5 to 2.2% from a 3 to 1 day HRT, respectively.

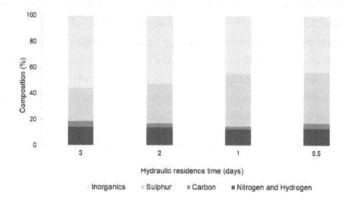

Figure 4.18: Effect of HRT on the elemental composition of the floating sulphur biofilm, showing the composition (%) of inorganics, sulphur, carbon, nitrogen and hydrogen present in the harvested FSB collected from the respective experimental run.

These findings suggested that the time between biofilm disruption and harvesting events affected the composition of the biofilm, with the ratio of elemental sulphur to organic material shifting. This is consistent with previous studies conducted by Mooruth et al. (2013) that revealed the relationship of FSB content (sulphur and organic material) as a function of HRT. The study concluded that a decrease in HRT led to an increase in the relative proportion of elemental sulphur while the organic carbon fraction decreased.

A large portion of the biofilm was made up of unknown inorganic compounds. Further investigation, through SEM-EDS analysis (Figure 4.19), detected large crystalline structures embedded within the biofilm, comprising of magnesium and phosphorus.

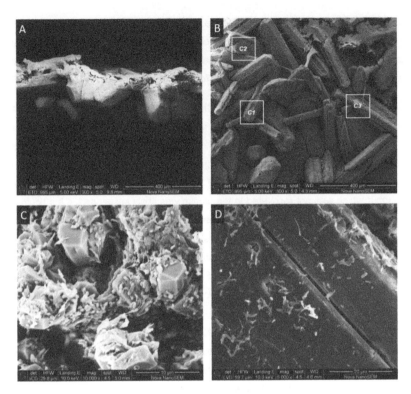

Figure 4.19: Scanning electron microscopy images of the inorganic crystals within the FSB showing A) side cross-sectional view, B) top view with biomass covering the surface and SEM-EDS elemental composition of the crystals performed in triplicate highlighted by areas C1-C3 detected mean composition containing C:28% O:40% Mg:13% P:17% (all elements were normalised), C) 10,000X view of biofilm covering smaller crystals, and D) 5,000X view of cells attached onto larger crystal.

The high inorganic content that accumulated in the biofilm hindered the maximum potential for sulphur recovery in hybrid system. Previous studies by Mooruth, (2013) reported biofilm compositions of 52 to 95% sulphur at a 4 and 2 day HRT, respectively, with inorganics accounting for less than 1%. By comparison, the sulphur recovery from the FSB in the hybrid LFCR was less efficient than when the LFCR was operated exclusively as a sulphide oxidation unit. However, this can primarily be attributed to the build-up of inorganics in the FSB as a result of the modified Postgate B medium (MPB) used in this study. MPB has been historically applied as a nutrient rich medium for the cultivation of SRB and to evaluate sulphate reduction (Postgate, 1979). However, it does contain excess amounts of magnesium and phosphate.

The reactor configuration and composition of the feed created conditions favouring the formation of the inorganic crystals in the biofilm, most likely as a result of evaporative crystallisation, where the floating biofilm at the air-liquid interface served as a nucleation site.

Though not to the same extent, the presence of inorganic crystals was reported by Mooruth (2013). Additional research is required to characterise the crystals and to elucidate the mechanism of its formation. The mass fraction of sulphur in the biofilm could be increased by adjusting the feed composition to reduce the formation of inorganic crystals within the FSB.

The hybrid LFCR was able to support high sulphide conversion, between 70 and 82% across the HRT range. The sulphate concentration in the final effluent was only marginally higher than in the reactor and no thiosulphate was detected, suggesting complete sulphide oxidation to sulphate was negligible. Liberation of H_2S gas or the formation of polysulphides was highly unlikely under the operating conditions (pH >7 and no turbulence) (Mooruth, 2013). Therefore, the sulphur which was not recovered in the biofilm was likely suspended in the liquid phase, either as colloidal sulphur or biofilm fragments. This was supported by the detection of high sulphur concentration through HPLC analysis and accumulation of biofilm fragments which settled in the effluent reservoir. The elemental sulphur balance analysis revealed that S^0 accounted for approximately 51% at a 3 day HRT and decreased to 14% at a 12 h HRT. The fraction of elemental sulphur in the effluent decreased with decreasing HRT. Cumulatively, the extended operation at a 3 day HRT (18 day operation) may have facilitated a higher release of sulphur into the effluent over time than the 12h HRT (3 day operation). Disrupting the biofilm mid-way through each HRT may have released elemental sulphur particles into the liquid phase as the broken fragments of biofilm settled onto the mesh-screen. Biologically produced sulphur is hydrophilic and dispersible in water, contrary to strongly hydrophobic and poorly soluble 'inorganic' elemental sulphur (S_8) (Kleinjan et al., 2005, Cai et al., 2017). This may facilitate its dispersion into the effluent stream.

4.5 Effect of biofilm disruption on process performance

In Sections 4.3 and 4.4, process performance of the hybrid LFCR was governed by a recurring trend in dissolved sulphide concentration, before and after biofilm disruption. In this study the effect of biofilm disruption is assessed over the 24 h window period after disrupting the biofilm. The study aims to better the understanding of periodically disrupting the biofilm on process performance, particularly on sulphate reduction.

4.5.1 Experimental approach

Following evaluation of HRT, the reactor was placed onto a 2 day HRT to re-establish optimal operation conditions. The commencement of the experiment began once stable system performance was obtained. The reactor was operated for three RTs after which the FSB was disrupted. Following disruption, the reactor performance was monitored at hourly intervals over 24 hours. The reactor was run for an additional three RTs before the biofilm was harvested.

4.5.2 Results and Discussion.

Disruption of the biofilm occurred after 7 day of operation. The removal of the biofilm resulted in a rapid decrease in dissolved sulphide concentration, from 4.5 mmol/L to 2.5 mmol/L within 12 hours, after which the concentration stabilised. A similar trend was observed in the effluent. Approximately 20 hours after disruption, the aqueous sulphide concentration began to increase, corresponding with re-forming the surface biofilm. After 24 hours a distinct thin layer biofilm was observed (Figure 4.8). The rapid decrease in dissolved sulphide concentration observed after biofilm disruption can be attributed to the increased oxygen mass transfer into the bulk liquid in the absence of the biofilm. The oxygen was rapidly consumed through the oxidation of sulphide. As the biofilm re-forms and matures, oxygen mass transfer into the bulk liquid was impeded and the rate of sulphide generation exceeded the sulphide oxidation rate, resulting in increased aqueous sulphide (Figure 4.20 A). Critically, the residual sulphate concentration remained stable during this period (Figure 4.20 B) indicating that sulphate reduction was not adversely affected during the 24 h period after biofilm disruption. It became evident that with sufficient residual aqueous sulphide to react with all the oxygen, it is possible to maintain anoxic conditions in the bulk volume. This was confirmed by redox potential measurements.

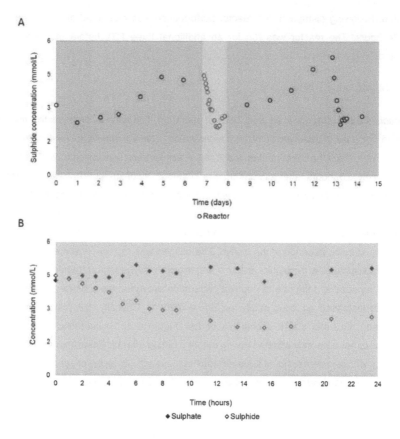

Figure 4.20: Effect of biofilm disruption on the performance of the hybrid LFCR showing A) mean reactor samples (FM, BM, BM and BB) over 15 days (24 h window period after biofilm disruption highlighted in grey) and B) sulphate and sulphide concentrations in the bulk volume for the 24 h after FSB disruption

The decrease in sulphide concentration was predominantly attributed to sulphide oxidation rather than the evolution of H_2S gas. This was consistent with the data from Mooruth (2013) who quantified H_2S (g) liberation to account for mass balance discrepancies in a similar system. The study concluded that the liberation of H_2S (g) from the LFCR surface was negligible. Most of the sulphide is present as HS^- at the operating pH and the lack of turbulent mixing reduces gas mass transfer across the surface. This is important from an aesthetic and safety perspective, due to the smell and toxicity of hydrogen sulphide gas.

Sulphide oxidation, with the generation of partially oxidised sulphur species such as colloidal sulphur, thiosulphate, polysulphides, or complete oxidation to sulphate may occur abiotically, or catalysed by microbes. For abiotic oxidation, the thermodynamics associated with the initial transfer of an electron for sulphide and oxygen reveal that the reaction is unfavourable, as an unstable superoxide and bisulphide radical ion would need to be produced (Luther et al., 2011). Alternatively, a two-electron transfer is favourable, with the formation of a stable S^0 and peroxide. However, the partially filled orbitals in oxygen that accept electrons prevent rapid kinetics. Due to these constraints the abiotic oxidation of sulphide is relatively slow. Alternatively, biologically mediated sulphide oxidation by photolithotrophic and chemolithotrophic microbes rely on enzymes that have evolved to overcome these kinetic constraints, allowing rapid sulphide oxidation. A study by Luther et al. (2011) demonstrated that biologically mediated sulphide oxidation rates are three or more orders of magnitude higher than abiotic rates. Furthermore, Mooruth (2013) investigated the extent of abiotic and biotic sulphide oxidation in a LFCR similar to that used in the current study. The study revealed that abiotic sulphide oxidation contributed little to the overall oxidation rate measured. This was based on a combination of kinetic constraints, poor oxygen diffusion, mixing (hydrodynamics) and convective mass transport within the LFCR. The biologically mediated oxidation of sulphide in the bulk volume following biofilm harvesting is critical. If the process relied on slower, abiotic oxidation, it is likely that the oxygen concentration in the bulk volume would increase to the point where sulphate reduction was inhibited.

As the biofilm develops, oxygen mass transfer into the bulk liquid is impeded and sulphide oxidation occurs exclusively within the biofilm. As the biofilm continues to mature and thicken, oxygen penetration through the biofilm slows to the point where it becomes limiting, resulting in a reduction in partial sulphide oxidation and an increase in dissolved sulphide concentration in the effluent, which is undesirable. Therefore, there is a need to optimise the frequency of FSB harvesting to ensure maximum sulphur recovery and consistent sulphide conversion.

4.6 Conceptual model of the hybrid LFCR process

The results discussed in Sections 4.3-4.5 demonstrated that efficient biological sulphate reduction and partial oxidation of sulphide to elemental sulphur can be achieved in a single reactor exposed to the atmosphere. The reactor maintained high sulphate reduction rates equivalent to those previously obtained in active, stirred tank reactors using a simple reactor geometry and requiring no energy input. The findings from this study resulted in the development of a conceptual model of the hybrid LFCR process, presented in Figure 4.21.

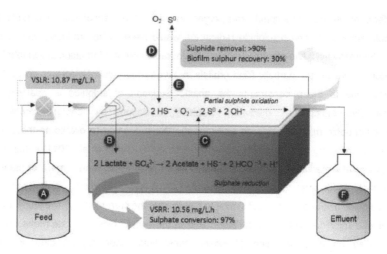

Figure 4.21: Conceptual model of the hybrid LFCR process and initial performance during demonstration at a 4 day HRT showing the configuration of the reactor setup, where a defined medium (A) is pumped into the reactor, (B) biological sulphate reduction occurs within the bulk volume and the generated sulphide (C) is partially oxidised at the air-liquid interface where oxygen (D) is available from the surrounding environment resulting in the formation of a floating sulphur biofilm. The elemental sulphur deposited within the biofilm (E) can be recovered through harvesting with the final treated effluent (F) characterised by low residual sulphate, sulphide and COD concentrations.

Feed solution (A), containing sulphate, sufficient organic carbon to sustain sulphate reduction and inorganic media components is pumped into the reactor at a defined hydraulic residence time. Biological sulphate reduction occurs within the anaerobic bulk volume of the reactor (B). Microbial attachment and subsequent colonisation of the carbon fibres facilitates biomass retention and an increased sulphate reduction rate. The absence of turbulent mixing ensures limited loss of gaseous hydrogen sulphide, ensuring good odour control, while the fluid flow pattern ensures delivery of sulphide to the biofilm. The FSB is formed, initially by heterotrophic species which produce an extracellular carbon matrix at the air-liquid interface. Autotrophic sulphur oxidisers colonise the biofilm. The biofilm results in an oxygen concentration gradient, creating a zone where the pH and redox environment favours microbially catalysed partial oxidation of sulphide to elemental sulphur. As the biofilm thickness and sulphur deposition increases, oxygen mass transfer is impeded to the point where the sulphide oxidation rate becomes slower than the sulphide generation rate and the sulphide concentration in the effluent increases. At this point, biofilm disruption or harvesting is necessary. Disrupting the biofilm removes the barrier to oxygen mass transfer so oxygen from the atmosphere (D) diffuses into the bulk volume, where it is used by planktonic SOB to oxidise sulphide within the bulk volume. This leads to a rapid decrease in sulphide concentration, but critically ensures that the bulk remains anoxic and sulphate reduction is not inhibited. The biofilm begins to re-form almost immediately and within 24 hours oxygen mass transfer is reduced to the point where the aqueous sulphide concentration increases again. By removing the harvesting screen situated just below the air-liquid interface the sulphur-rich biofilm can be harvested and elemental sulphur (E) recovered as a value added product. The final treated effluent (F) discharged at the outlet port at the liquid surface is characterised by low residual COD, sulphate and sulphide concentration with an elevated pH.

4.7 Conclusions

The hydrodynamics in the 2 L LFCR configuration used for the demonstration of the hybrid process were shown to be governed by a mixing regime consistent with that previously shown (Mooruth, 2013) within a 25 L LFCR configuration. Though simple in design and governed solely through passive mixing, the LFCR exhibited a dynamic fluid mixing pattern and was relatively well-mixed as a result of high diffusive mixing.

The successful demonstration of the hybrid LFCR, with the integration of sulphate reduction and partial sulphide oxidation within a single reactor unit, achieved near complete sulphate conversion (97%) to sulphide with partial sulphide oxidation under the conditions tested, leading to elemental sulphur recovery. Biofilm disruption and harvesting was an important parameter in successfully operating the hybrid LFCR. The correct regulation of the biofilm ensured optimal process performance in terms of sulphate reduction and partial sulphide oxidation via the FSB. The disruption of the biofilm did not adversely affect sulphate reduction performance, provided sufficient sulphide was present within the bulk volume prior to biofilm disruption.

The outcome of the effect of HRT study confirmed that HRT plays a critical role in the overall performance of the hybrid LFCR. High rates of biological sulphate reduction and partial sulphide oxidation were achieved over the range of HRTs, with a portion of the sulphur recovered through harvesting the biofilm. Biomass retention, by attachment to carbon microfibers, supported high kinetic rates, even at the highest dilution rate of 12 h. This demonstrated the effectiveness of carbon fibres as a support matrix. The VSRR was positively correlated to HRT, increasing proportionally as HRT decreased across the experimental range (3 – 0.5 day HRT). While the increase in VSRR was sustained even at a 1 day and 12 h HRT, the sulphate conversion was negatively affected. The competitive interaction of lactate fermenters and oxidisers (SRB) influenced the system performance where a metabolic shift in lactate utilisation was observed at high dilution rates. Therefore, optimal system performance as a function of HRT was characterised by a compromise between VSRR and sulphate oxidation efficiency. Based on overall performance, operation at a 2 day HRT was selected as optimal. Sulphide oxidation kinetics relied on the available sulphide generated as a consequence of sulphate reduction, as a result the degree of VSOR correlated with the VSRR. The importance of regulating biofilm disruption and harvesting in the hybrid LFCR, to maximise sulphur recovery, was demonstrated.

The successful demonstration of the hybrid process generated interest to evaluate the technology at pilot scale, necessitating further development of the process with a focus on evaluating key challenges that would be experienced at a larger scale. In the following chapters the effects of key operational parameters on the process performance are investigated towards further characterisation of the process.

Chapter 5

Reactor scale-up and geometry

5.1 Introduction

Following the successful demonstration of the hybrid LFCR process at laboratory-scale, the promising results and potential application of the process to treat sulphate-rich waste streams generated interest to evaluate the technology at larger scale and led to the commissioning of the pilot plant. To ensure successful operation of the pilot plant, further process development was needed to evaluate key potential challenges at a larger scale.

The pilot-scale reactor (2025 L) was designed based on the 2 L LFCR, described in Section 3.1.1. However, due to construction constraints using Plexiglass material, the aspect ratio of the pilot-plant reactors was slightly altered. This resulted in the decision to construct a laboratory-scale reactor that simulated the pilot plant dimensions.

The aim of this investigation was to evaluate the impact of the altered aspect ratio and upscaling of the reactor volume, from 2 L to 8 L, on the system performance of the hybrid LFCR process. In addition, the effect of HRT as a key operational parameter across both systems was evaluated.

The specific objectives addressed are as follows:

1. Determine the hydrodynamic regime within the 8 L LFCR and assess whether it conforms to the 2 L LFCR.
2. Evaluate the effect of scale-up and altered reactor geometry on process performance by comparing these LFCRs.
3. Assess the effect of HRT on process performance as a function of reactor scale-up and design.
4. Evaluate the effect of biofilm disruption regime, as a function of operating time, on process performance of the 8 L LFCRs.

5.2 Reactor design of the 8 L LFCR

The upscale laboratory LFCR reactor was designed to simulate the geometry of the pilot plant. It was constructed from Plexiglass (10 mm), with an internal dimension of 450 mm (l) x 200 mm (w) x 150 mm (h) (Figure 5.1).

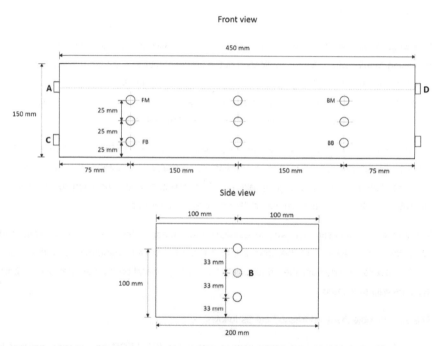

Figure 5.1: Schematic diagram of the 8 L LFCR simulating the dimensions of the pilot scale in aspect ratio showing the front and side view and the location of A) inlet port B) carbon microfibers C) heat exchanger and D) outlet port.

Operation at 100 mm liquid height resulted in a working volume of approximately 8 L. As with the original 2 L LFCR, the 8 L variant was fitted with sampling ports on the front facing reactor wall, across the reactor length and depth. The left and right side reactor walls each contained three tapped (1/8" BSP) holes. The uppermost holes represented the feed and effluent ports (Figure 5.1; A and D), while the lower holes (Figure 5.1; C) were used as attachment points for the heat exchanger. The middle holes were blind holes (Figure 5.1; B) which supported the rod, fitted with carbon microfibers, in position. The harvesting screen was held in position, just below the air-liquid interface, using wire hooks which were bent around the side walls of the reactor (Figure 5.2).

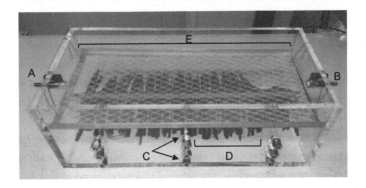

Figure 5.2: Photograph of the 8 L LFCR, features of the reactor that are shown include A) inlet port, B) outlet port C) sampling ports D) carbon microfibers E) harvesting screen.

5.3 Hydrodynamics

5.3.1 Experimental approach

In Section 4.2, the hydrodynamics in the 2 L LFCR, characterised by a visual tracer study, confirmed that the mixing regime was consistent with the conceptual model described by Mooruth (2013). The outcome of a similar investigation in the 8 L design would inform any change that may impact the overall performance of the system and confirm whether the difference in aspect ratio impacted the overall hydrodynamic mixing profile. If negligible, comparative experimental studies between the 2 L and 8 L configurations could be performed, where the influence of the hydrodynamics would be negligible across the two reactors configurations.

The comparative dye tracer study (as described in Section 3.4) was performed in both 2 L and 8 L LFCRs. This included an assessment of the effect of HRT and whether hydrodynamics was preserved across changes in geometry. The study tested a range of HRT conditions (4, 3, 2, 1 and 0.5 day) and was performed in triplicate.

5.3.2 Results and discussion

The hydrodynamic profile analysis of the 8 L LFCR, based on visual tracer studies, was similar to that obtained within the 2 L LFCR, despite the difference in aspect ratio (Figure 5.3). The flow pattern was consistent with that reported in Section 4.2 and was governed primarily by advective and diffusive mass transport with limited turbulent mixing.

The 8 L reactor did, however, exhibit subtle differences in mixing compared to the 2 L configuration. In the 2 L LFCR, an initial zone of clearing ran across the base of the reactor, which mimicked elements of parabolic laminar flow. This zone of clearing reached the outlet end of the reactor well before decolouration of the bulk volume occurred. The distinctive pattern corresponded well with findings obtained in Section 4.2 as well as with the conceptual model (Figure 4.2) described by Mooruth (2013). In the larger 8 L reactor, more diffusive mixing was observed during the initial stages of the experimental run (Figure 5.3 A). This was primarily attributed to the higher flow rate required to achieve an equivalent HRT to the 2 L LFCR. As a result, a higher fluid velocity led to the increased formation of turbulent eddies upon entering the reactor, increasing the rate of diffusive mixing near the inlet port.

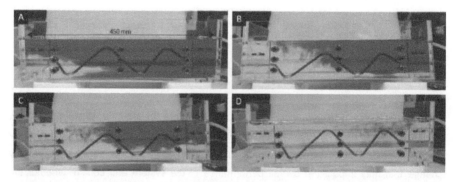

Figure 5.3: Photographic images showing the progression of mixing in the 8 L LFCR configuration at a 2 day HRT at ambient temperature, photographs taken at 54 min (A), 70 min (B), 85 min (C) and 98 min (D).

The fluid dynamic pattern was consistent across the range of HRTs tested (4 – 0.5 day HRT), with minimal differences observed with repeat experiments. The decrease in operating HRT was followed by a decrease in the complete mixing times (Figure 5.4). These remained substantially shorter than the overall HRT at which the reactor was operated. This was consistent with the initial dye tracer study performed in Section 4.2.

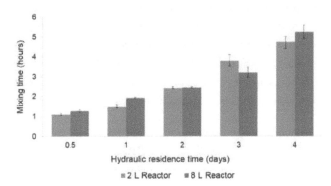

Figure 5.4: Complete mixing times as a function of HRT for the 2 and 8 L reactors, represented by an average of three replicates. Error bars represent the standard deviation observed between experimental runs.

The variation in reactor configuration in terms of aspect ratio and scale-up had minimal effect on the overall fluid mixing dynamics with both configurations exhibiting similar complete mixing times across the HRT range.

5.4 Effect of hydraulic residence time

5.4.1 Experimental approach

5.4.1.1 Adjustment to operating condition and maintenance of the 2 L LFCR

Following the successful demonstration and evaluation of the effect of hydraulic residence time on process performance (Chapter 4), the results highlighted that the exposure to a 12 h HRT caused an increase in fermentation of lactate. This was particularly evident by the loss in overall sulphate reduction performance and concomitant increase in propionate concentration. This likely resulted in a shift in the microbial community structure, favouring the proliferation of fermentative microorganisms. By the end of the HRT study, in order to re-establish optimal performance, the system was placed on a 2 day HRT. Soon after, an investigation into the effect of biofilm disruption over 24 h was initiated. However, the sulphate reduction performance previously obtained within the system could not be recovered and performance stabilised at approximately 50% removal. The extent of exposure to high dilution rates below a 1 day HRT is expected to affect the microbial community structure, consequently resulting in a decrease in process performance.

After evaluating the effects of biofilm disruption, the 2 L LFCR was operated continuously with regular biofilm disruption and harvesting up until day 649, from the time of start-up. Over this period, attempts to recover system performance were initiated by increasing the HRT to 4 days. In addition, the operational temperature was increased to 30°C to promote higher sulphate reducing activity. It was anticipated that the longer HRT would favour the growth of SRB over fermentative bacteria by facilitating substrate limiting conditions, while increasing the operating temperature at 30°C would increase sulphate reduction activity.

Section 4.4.2 showed that sulphur recovery via the FSB was limited due to the formation of large inorganic crystals that were embedded within the biofilm. Upon further analysis using SEM-EDS it was revealed that the inorganic crystals were predominantly comprised of magnesium, phosphorus and nitrogen. This was attributed to the feed composition which comprised of magnesium, phosphate and ammonia. Subsequently, the feed composition (Section 3.1.2) was modified to reduce the formation of the inorganic precipitate with the aim of increasing sulphur recovery in the biofilm. $MgSO_4.7H_2O$, as the main source of sulphate, was reduced from 2 to 1 g/L and $NaSO_4.2H_2O$ adjusted from 0.3 to 0.9 g/L to ensure the feed sulphate concentration of 1 g/L.

During the initial demonstration (Chapter 4), the effluent was sampled at the end of the silicone effluent pipe that fed into an effluent reservoir. However, a considerable amount of sulphur built up within the pipe over time, often resulting in blockage. Furthermore, the effluent sample contained a large portion of particulate matter resembling elemental sulphur and fragments of FSB. Upon further analysis of the effluent samples, low sulphide concentrations (± 2 mmol/L) were consistently detected regardless of whether high sulphide concentrations were measured in the samples drawn directly from the reactor. This was attributed to sulphide oxidation of the effluent in the silicone pipe and was supported by an investigation which evaluated the potential application of sulphide oxidation in a silicone tubular reactor (Rein 2002). The oxygen permeability of silicone tubing promoted suitable oxygen limiting conditions and facilitated partial sulphide oxidation to elemental sulphur. As a consequence, in the current investigation, the sampling procedure of the effluent was altered to ensure that the effluent sample reflected the performance of sulphide oxidation within the reactor through minimising additional oxidation downstream. The effluent pipe was detached and effluent overflow collected directly at the outlet port.

Together, these adjustments were necessary to recover optimal process performance and to address key issues identified during the initial demonstration of the hybrid LFCR process.

5.4.1.2 Start-up procedure of the 8 L LFCR

An 8 L LFCR was set up and inoculated with a batch inoculum collected from the overflow from the original 2 L LFCR. The decision to inoculate the reactor with culture derived from the 2 L LFCR was to ensure that any disparity in system performance and microbial community structure obtained in the 8 L LFCR variant would be more likely a function of reactor configuration than variation in the microbial ecology of the start-up inoculum. The overflow was initially maintained as a batch culture at 30°C in a closed vessel on a magnetic stirrer to build-up of sufficient volume prior to inoculation of the 8 L reactor. Regular sub-culturing promoted the development of a highly active sulphate reducing culture. A 50% (v/v) subculture was conducted weekly, using the same synthetic feed fed into the LFCR. A critical requirement was to ensure sufficient dissolved sulphide concentration upon start-up to promote the rapid development of the FSB as well as establish the anaerobic environment in the bulk volume required for SRB growth.

The 8 L LFCR was set-up alongside the 2 L LFCR and operated at the same conditions described in Section 5.4.1.1 i.e. at a 4 day HRT and 30°C initially. Furthermore, the altered feed composition was used to supply both the 2 L and 8 L LFCR.

5.4.1.3 Effect of hydraulic residence time

The investigation into the effect of HRT on process performance as a function of reactor design was performed similarly to the experimental approach adopted in Section 4.4. The study was initiated once the 8 L LFCR established stable performance, achieved after the initial start-up and acclimatisation period. A range of HRTs (4, 3, and 2 day) were tested by changing the feed rate into the 2 and 8 L LFCR reactors. In this investigation the maximum dilution rate was set at a 2 day HRT, shown to be optimal in the 2 L LFCR in Section 4.4. At the beginning of each HRT condition, a biofilm harvest was performed and the reactors were operated continuously with routine monitoring of system performance. After a set time period, biofilm disruption was performed. The reactors proceeded for an additional period, followed by a second biofilm disruption and harvest event. After harvesting the biofilm, the feed rate was increased to achieve the next HRT and the process repeated.

5.4.1.4 Effect of biofilm disruption regime

In addition to assessing the effect of HRT across reactor configurations, an experiment was carried out to evaluate the effect of biofilm disruption regime as a function of time. While Section 4.5 highlighted the effects of biofilm disruption on process performance over 24 h after disruption, the objective of this study was to determine the impact of different time intervals of operation between biofilm disruption events. This was performed exclusively on the 2 L LFCR at the beginning of the current study, alongside the start-up of the 8 L LFCR. The investigation involved two experimental runs operated at a 4 day HRT, denoted as experiment 4(1) and 4(2). Experiment 4(1) followed the strict biofilm disruption and harvest regime adopted in Section 4.4 and involved disrupting the biofilm after 3 HRTs (12 days) followed by a subsequent harvest after an additional 3 HRTs of operation. In contrast, during experimental run 4(2), the system was operated for an extended period with a biofilm disruption only occurring after 29 days (approximately 7 HRTs).

5.4.2 Results and discussion

5.4.2.1 Effect of reactor geometry on process performance

Residual sulphate concentration measured over the duration of the study for both 2 L and 8 L LFCR is presented in Figure 5.5. Minimal variation in sulphate concentration was observed across reactor sampling ports (FM, FB, BM and BB) which demonstrated the relatively well mixed environment present within the reactors.

During the initial experimental run 4(1) conducted in the 2 L LFCR (Figure 5.5 A), under strict biofilm disruption regime, the residual sulphate concentration decreased to approximately 6 mmol/L after 3 residence times of operation. Following a biofilm disruption, the sulphate increased before gradually decreasing over 3 residence times to a similar concentration prior to disruption. Alternatively, by applying an extended period of operation (29 days) in experimental run 4(2), the sulphate concentrations sharply decreased to 3.8 mmol/L. The results show that longer operation between FSB disruption and harvest promoted higher sulphate conversion to be achieved. During start-up of the 8 L reactor, it was anticipated that at a larger scale the system would exhibit a longer lag and acclimatisation phase. However, after just 27 days of continuous operation, the 8 L LFCR reached stable performance similar to that of the 2 L LFCR, based on residual sulphate concentration. The formation of the FSB, within 24 h after inoculation, at the air-liquid interface (Figure 5.6) aided in the development of anoxic conditions in the bulk volume by impeding oxygen penetration, providing a suitable environment for SRB growth.

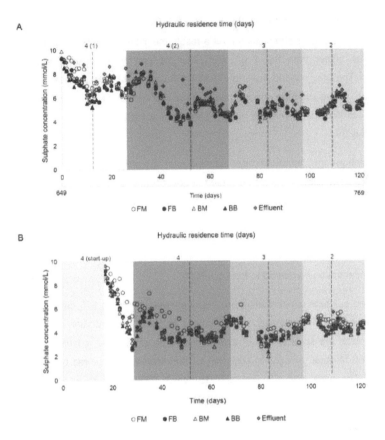

Figure 5.5: Residual sulphate concentration as a function of sample ports and hydraulic residence time of the 2 L (A) and 8 L LFCR (B). Biofilm disruption and harvest events are indicated by vertical dotted and solid lines, respectively. A change in HRT is indicated by the transition in shading intensity and was accompanied by a biofilm harvest.

Figure 5.6: Photograph of the 8 L LFCR, showing the development of a floating sulphur biofilm at the air-liquid interface within the 24 h after initial inoculation.

Prior to inoculation of the 8 L LFCR, there was an accumulation of sulphate in the inoculum, a consequence of regular sub-culturing. As a result, the initial sulphate concentration of the inoculum was higher than the 1 g/L feed sulphate concentration used in the reactor experiments. The grey shaded area, in Figure 5.5 B, represents the unstable period during initial start-up, where sulphate concentrations measured >10 mmol/L. From day 17, residual sulphate concentrations decreased rapidly over time from approximately 10 mmol/L to 2.9 mmol/L by day 29.

Residual sulphate concentrations were affected after a biofilm disruption or harvest event (Figure 5.5). Both systems were observed to experience a brief increase in sulphate concentration before gradually decreasing to a similar concentration achieved prior to the disruption. In evaluating the effect of HRT, both reactors displayed a similar residual sulphate concentration profile over time with a gradual increase in the pseudo-steady state concentration just before biofilm disruption. During the study, the 8 L LFCR achieved lower sulphate concentration compared to the 2 L LFCR.

The dissolved sulphide concentration profiles, presented in Figure 5.7, displayed a similar trend to Section 4.4. This involved an increase in sulphide concentration over time as the biofilm formed at the surface with a rapid decrease observed after a biofilm disruption or harvest event. Minimal variation in sulphide concentration was observed between different reactor sampling ports (FM, FB, BM, and BB) which corresponded well with the residual sulphate data (Figure 5.5).

The effluent concentration profile followed a similar trend observed within the reactor samples with increasing and decreasing concentrations as a function of biofilm disruption or harvest event. Throughout the study the effluent concentrations were on average lower than the reactor samples and ranged between 1 – 7 mmol/L. This was distinctly different to the results obtained in Section 4.4, where the effluent sulphide concentrations were consistently low at ±2 mmol/L, across the range of HRT evaluated. This was attributed to the change in sampling procedure that was adopted in the current investigation to obtain reliable measurement of sulphide exiting the reactor (Figure 5.7).

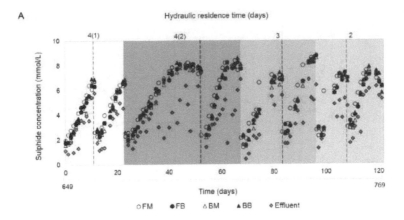

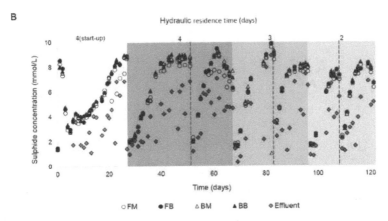

Figure 5.7: The effect of HRT on the dissolved sulphide concentration measured in the reactor (FM, FB, BM and BB) and effluent samples over time in the A) 2 L and B) 8 L lactate-fed reactors. Biofilm disruption and harvest events are indicated by vertical dotted and solid lines, respectively. A change in HRT is indicated by the transition in shading intensity and was accompanied by a biofilm harvest.

The mean pH data obtained from both 2 L and 8 L LFCR is shown in Figure 5.8. A consistent trend was observed across both reactor configurations, with a sharp increase in the average pH measured in the reactor samples, after a biofilm disruption or harvest event. The pH gradually decreased towards a pH range equivalent to the feed (pH 7). The observed trend in pH coincided with the observed sulphide concentration profile in Figure 5.7.

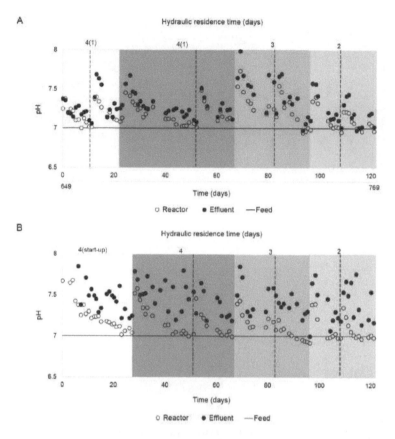

Figure 5.8: Mean pH data of reactor samples (FM, FB, BM, BB) and effluent as a function of hydraulic residence time in the A) 2 L LFCR (blue) B) 8 L LFCR (purple). Biofilm disruption and harvest events are indicated by vertical dotted and solid lines, respectively. A change in HRT is indicated by the transition in shading intensity and was accompanied by a biofilm harvest.

After biofilm disruption a rapid decrease in sulphide was accompanied by an increase in pH. This was attributed to partial sulphide oxidation (Reaction 2.18), during which hydroxyl ions are released. On average the pH measured in the effluent, from both reactors, was consistently higher than the reactor samples. Further, the effluent pH in the 8 L LFCR was higher and more variable compared to the 2 L LFCR. The release of the effluent from the reactor was governed by gravitational flow and the flow rate within the 8 L LFCR was more variable than in the 2 LFCR. This was consistent with the observed variation in effluent pH and sulphide concentration in the 8 L LFCR.

VFA analysis was conducted to evaluate overall lactate utilisation as a function of HRT. Lactate metabolism can proceed via oxidation (Reaction 2.12 and 2.13) or fermentation

(Reaction 2.15) to acetate and propionate. In the VFA concentration profiles of lactate, propionate and acetate, shown in Figure 5.9, complete utilisation of lactate was observed, with residual lactate concentrations below the detection limit. The high acetate and corresponding residual sulphate concentration indicated that incomplete oxidation of lactate by SRB dominated both reactors.

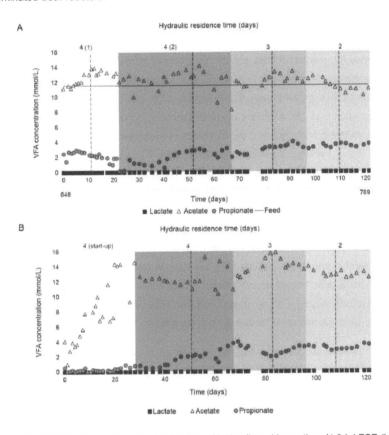

Figure 5.9: Volatile fatty acid profile as a function of hydraulic residence time A) 2 L LFCR (blue) B) 8 L LFCR (purple), data represents mean values from reactor sampling ports (FM, FB, BM, and BB). Biofilm disruption and harvest events are indicated by vertical dotted and solid lines, respectively. A change in HRT is indicated by the transition in shading intensity and was accompanied by a biofilm harvest.

In the 2 L LFCR, measured acetate concentrations were relatively stable, ranging between 11 and 14 mmol/L (Figure 5.9 A). The increasing and decreasing trends of acetate utilisation coincided with the biofilm disruption and harvesting events, like that observed in the residual sulphate concentration (Figure 5.5).

Collectively this suggests that sulphate reduction activity was briefly affected by biofilm disruption, most likely due to the influx of oxygen through the surface, which may have briefly inhibited SRB activity. A relatively stable amount of propionate was produced in the 2 L LFCR (Figure 5.9 A). During experimental run 4(1), propionate concentration measured at 3 mmol/L. There was a notable decrease in propionate concentration to approximately 1 mmol/L during experimental run 4(2) which was followed by a gradual increase as the HRT was decreased to a 2 day HRT, reaching a maximum of 3.8 mmol/L.

Similarly, in the 8 L reactor, acetate concentration rapidly increased, from 1 to 14.5 mmol/L, during the initial start-up period. The increase in acetate accumulation and concomitant decrease in residual sulphate (Figure 5.5 B), indicated that incomplete oxidation of lactate for sulphate reduction was the dominant metabolic pathway occurring within the system. On average the acetate concentrations in the 8 L reactor ranged between 11 and 15 mmol/L which was slightly higher than that observed in the 2 L reactor. This was consistent with the higher sulphate conversion and sulphide concentration observed in the 8 L reactor. Minimal propionate was produced during the initial start-up of the 8 L reactor. However, while operated at a 4 day HRT, a sudden increase in propionate concentration occurred on day 38 (Figure 5.9 B), indicating an increase in lactate fermentation. While the fermentation of lactate to acetate and propionate in BSR systems has been well documented (Bertolino et al., 2012: Oyekola et al., 2012), most studies have reported the increase in fermentation only occurring once operated under excess lactate concentrations or high dilution rates. This is due to lactate fermenters typically having a higher growth rate, but lower affinity for lactate than SRB and therefore are outcompeted under carbon limiting conditions.

These findings were supported by the results obtained in Section 4.4, where a decline in sulphate conversion was accompanied by the increase in propionate concentration once operated at high dilution rates equivalent to a 1 and 0.5 day HRT. After evaluating the performance at a 0.5 day HRT, attempts to recovery performance in the 2 L LFCR were initiated by increasing the HRT to 4 days. The system was continuously operated for 446 days, during which stable production of propionate was observed within the system. This suggested that even though the operating conditions were unfavourable for lactate fermentation, due to the configuration of the hybrid LFCR to retain biomass, a fermentative microbial population had accumulated and established within the system after the exposure to high dilution rates. Oyekola et al. (2009) reported that the increase in fermentation of lactate at high dilution rate was concomitant with a shift in the microbial community structure, in which an increase in fermentative microorganisms was observed. Since the 8 L LFCR start-up inoculum was derived from the 2 L LFCR, the similar performance observed across both reactors including the proportion of lactate fermentation, which was consistently maintained, may be function of the microbial community composition.

During the investigation after 40 days of operation, propionate concentration on average ranged between 3 and 3.7 mmol/L in the 2 and 8 L LFCRs, respectively. If the portion of lactate lost to fermentation, based on the measured propionate concentration, was directed towards sulphate reduction, stoichiometrically, this would be equivalent to approximately 20-27% conversion of the feed sulphate concentration (1 g/L). This is substantial when considering the overall biological sulphate reduction performance and indicates that the process was limited by lactate fermentation.

The volumetric substrate utilisation and production rate profiles as a function of HRT are presented within Figure 5.10. The results reveal a linear increase in all volumetric rates associated within lactate metabolism and sulphate reduction as the HRT was decreased from 4 to 2 days (dilution rate 0.0104 – 0.0138 1/h) in both 2 L and 8 L reactors. There was a pronounced increase in both acetate and propionate production rate within the 8 L reactor which was accompanied by a higher sulphate reduction rate compared to that observed within the 2 L reactor. The total residual VFA measured (lactate, acetate and propionate concentrations) exceeded the theoretical value based on the feed lactate concentration, considering fermentation and oxidation reactions. The in-depth analysis of the stoichiometry of lactate metabolism and carbon balance is discussed in Section 5.4.2.2.

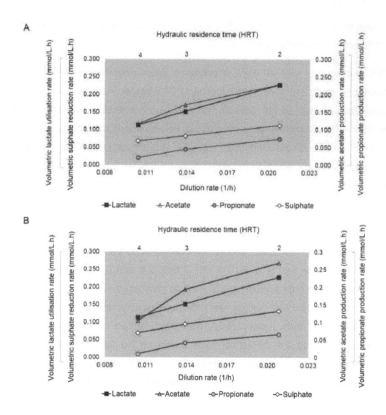

Figure 5.10: Effect of HRT on the volumetric sulphate reduction, substrate (lactate and acetate) utilisation and production (acetate and propionate) rates in the A) 2 L and B) 8 L lactate-fed reactors.

5.4.2.2 Stoichiometric dependence on hydraulic residence time and reactor configuration

From the results presented in Figure 5.10, lactate metabolism toward sulphate reduction was favoured in both reactors throughout the study. The range of HRT applied favoured SRB activity over fermentation. In this study, lactate was supplemented at a COD/SO$_4$ (g/g) ratio of 0.7, which is slightly higher than the theoretical ratio (0.67) to achieve 100% sulphate conversion through complete oxidation. However, sulphate reduction predominantly occurred via incomplete oxidation (Reaction 5.2) to acetate, hence only 53% sulphate conversion was theoretically possible using the incomplete oxidation pathway. Taking this into consideration (Table 5.1), the sulphate reduction performance in the reactors was lactate limited via partial oxidation of the substrate (Reaction 5.2), with the complete oxidation of lactate being the rate limiting step in the current investigation (Reaction 5.1).

Oyekola et al. (2010) evaluated lactate metabolism and sulphate reduction kinetics in a CSTR using lactate at 20% excess of the molar sulphate concentration to ensure that sulphate reduction through incomplete oxidation was not limited (Reaction 5.2). The study evaluated lactate metabolism under sulphidogenic conditions and assessed the effect of sulphate loading through dilution rate and feed sulphate concentration. It achieved high sulphate conversion and provided a detailed investigation into lactate metabolism and the competition between SRB and fermenters. However, for practical implementation, the addition of an electron donor to treat a sulphate-rich waste stream needs to be supplemented near the theoretical ratio to limit secondary pollution as a result of excess COD in the effluent. One of the major drawbacks of using lactate (Oyekola et al., 2010; Celis et al., 2013) and ethanol (Erasmus, 2000) is the accumulation or inefficient metabolism of acetate. In the current work, the rationale for operating the reactors at a ratio of 0.7 COD/SO$_4$ (g/g) was to evaluate lactate as a sole carbon source based on complete oxidation (Reaction 5.1). As a result, lactate was supplemented at <5% in excess of the theoretical COD required to treat the feed sulphate.

In the current investigation, based on the experimental data, the stoichiometric ratios (L:A, L:S and L:A) were determined and given in Table 5.2. A similar approach was applied by Erasmus (2000) and Oyekola et al. (2010), when evaluating ethanol and lactate metabolism, respectively, under sulphidogenic conditions. The experimental stoichiometric ratios L:A, L:S and A:S (Table 5.2) largely agreed with the theoretical values for incomplete oxidation (Reaction 5.2), given in Table 5.1, in both 2 L and 8 L reactors.

Table 5.1: Theoretical stoichiometric ratios of the metabolic reactions associated with the respective electron donor.

Reaction No.	Chemical reaction	Theoretical stoichiometric ratio		
		L:A	L:S	A:S
5.1	$2\ Lactate + 3\ SO_4^{2-} \rightarrow 6\ HCO_3^- + 3\ HS^- + H^+$	-	0.67	-
5.2	$2\ Lactate^- + SO_4^{2-} \rightarrow HS^- + 2\ Acetate^- + 2\ HCO_3^- + H^+$	1.0	2.0	2.0
5.3	$3\ Lactate \rightarrow acetate + 2\ propionate + HCO_3^- + H^+$	3.0	-	-

Table 5.2: Effect of HRT on the molar ratio of lactate utilised to moles of acetate produced involved in biological sulphate reduction, using lactate as the sole carbon-source and electron donor. Average values of experimental stoichiometric ratios are compared with the theoretical ratios (Table 5.1). The carbon balance of total moles VFA accounted compared with the total amount of lactate fed is also presented.

HRT (days)	Volumetric rates (mmol/L.h)				Stoichiometric ratios			Carbon balance
	Lactate utilisation rate	Acetate production rate	Propionate production rate	Sulphate reduction rate	Total moles lactate used/mole acetate produced (L:A)	Total moles lactate used/mole sulphate reduced (L:S)	Total Moles acetate produced/ mole sulphate reduced (A:S)	Total C moles out/ Total C mole lactate fed [b] (Effluent:Influent)
					2 L lactate-fed			
4	0.114	0.116	0.021	0.068	1.0	1.7	1.7	0.9
3	0.152	0.171	0.044	0.083	0.9	1.8	2.1	1.0
2	0.228	0.228	0.074	0.113	1.0	2.0	2.0	1.0
					8 L lactate-fed			
4	0.114	0.104	0.010	0.07	1.1	1.6	1.5	0.7
3	0.152	0.193	0.042	0.095	0.8	1.6	2.0	1.1
2	0.229	0.268	0.066	0.132	0.9	1.7	2.0	1.1

[b] Carbon balance of the total mol C measured to the total amount of mol C lactate fed

Assuming the generation of propionate proceeded via Reaction 5.3, the amount of lactate utilised and acetate generated via fermentation can be accounted for. By accounting for the contribution of lactate metabolism via fermentation, the resultant estimated experimental ratios provide a more accurate overview of the sulphate reduction stoichiometry occurring within the reactors based on lactate metabolism (Reaction 5.1 and 5.2). This had an impact on the stoichiometric profiles observed in Figure 5.11, with a decrease in all three experimental ratios L:A, L:S and A:S in both reactors. The most pronounced deviation was observed amongst the L:S ratio which indicated that less lactate was utilised for the observed sulphate reduction compared with the theoretical of incomplete (Reaction 5.2) lactate oxidation (Figure 5.11 B). Instead the experimental ratios (SR) diverged more toward the theoretical for complete (Reaction 5.1) lactate oxidation.

Overall stoichiometric ratios were consistent with the observed sulphate reduction performance observed in both reactors. Although sulphate reduction via incomplete oxidation of lactate was limited by availability of electron donor, the current study achieved 63 and 65% sulphate conversion during operation at a 4 day HRT in the 2 L and 8 L reactors, respectively. This demonstrated that complete oxidation of lactate can be stimulated under low dilution rates that favour acetate utilising SRB.

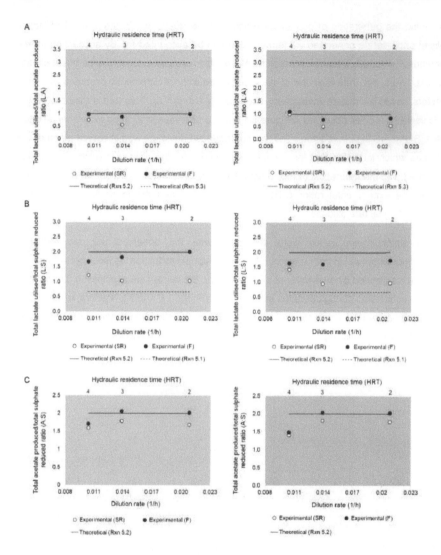

Figure 5.11: Effect of HRT on biological sulphate reduction stoichiometry in the 2 L (left) and 8 L (right) reactors, showing the A) Total moles of lactate utilised per mole total acetate produced (L:A), B) total moles of lactate utilised per total moles sulphate reduced (L:S), C) moles of acetate produced per total moles of sulphate reduced (A:S). Experimental ratio with (F) and without (SR) the contribution of fermentation, calculated stoichiometrically (Rxn 5.3) based on residual propionate concentration. The horizontal solid (Rxn 9.2) and dotted (Rxn 9.1 and 9.3) lines represent the theoretical ratio for the respective reactions.

5.4.2.3 Biological sulphate reduction kinetics

In the hybrid system, the expected sulphide, available for partial oxidation, is calculated based on the sulphate reduced. The sulphide converted through sulphide oxidation was determined by the difference between the theoretical sulphide expected and final dissolved sulphide measured in the effluent (Section 3.7.1; Equation 3.6). The expected sulphide produced and the measured sulphide concentrations in the reactor and effluent is shown in Figure 5.12. After biofilm disruption there is a rapid decline in sulphide concentration in the reactor and effluent. The effluent sulphide concentrations were consistently lower than the reactor samples, as expected since the effluent port is located at the surface, resulting in additional oxidation of the sulphide. As the FSB develops and becomes oxygen limiting, sulphide accumulates over time within the reactor and effluent samples. As a result, the difference between the expected sulphide and final effluent sulphide concentration becomes negligible. The results illustrate the high portion of sulphide removal that occurs directly after biofilm disruption.

Analysis of biological sulphate reduction kinetics revealed that under a strict biofilm disruption regime, during experimental run 4(1), a volumetric sulphate reduction rate (VSRR) of 0.048 mmol/L.h with a corresponding conversion of 44% was achieved. In contrast, by extending the duration between biofilm disruption events, in experimental run 4(2), the VSRR increased to 0.068 mmol/L.h with a conversion of 63%. Although regulating the biofilm more stringently facilitated a higher biofilm recovery to sulphide-S ratio, it limited the biological sulphate reduction performance.

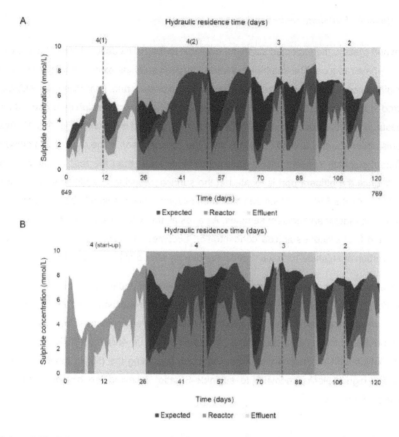

Figure 5.12: Sulphide concentration profile across the hybrid LFCR showing the expected sulphide generated, average sulphide concentration present in the reactor samples (FM, FB, BM, and BB) and final sulphide measured in the effluent as a function of hydraulic residence time in A) 2 L and B) 8 L lactate-fed reactor. Biofilm disruption and harvest events are indicated by vertical dotted and solid lines, respectively. A change in HRT is indicated by the transition in shading intensity and was accompanied by a biofilm harvest.

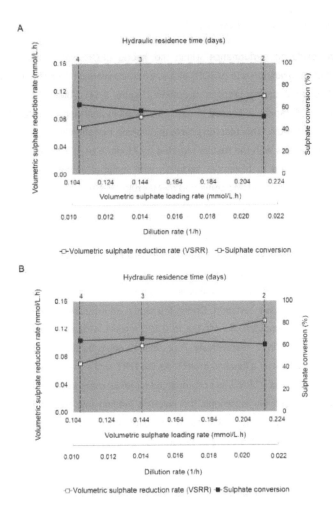

Figure 5.13: Pseudo-steady state kinetic data of volumetric sulphate reduction rates and sulphate conversion as a function of hydraulic residence time showing A) 2 L and B) 8 L lactate-fed reactors.

The stable performance based on sulphate reduction kinetics obtained during the HRT study is presented in Figure 5.13. On assessing the effect of HRT, similar trends were observed across both reactor configurations. A linear increase in VSRR (2 L: 0.068 – 0.113 mmol/L.h; 8 L: 0.070 – 0.132 mmol/L.h) was accompanied by a gradual decrease in sulphate conversion (2 L: 63 – 52%; 8 L: 65 – 61%) as the HRT was reduced from 4 to 2 days, respectively. Overall, the 8 L LFCR performed better than the 2 L LFCR, achieving the highest VSRR of 0.132 mmol/L at a 2 day HRT and highest sulphate conversion of 65% at a 4 day HRT.

5.4.2.4 Biological sulphide oxidation kinetics

The sulphide oxidation performance within the hybrid LFCR was evaluated based on volumetric sulphide oxidation rate (VSOR) and sulphide conversion. The VSOR and sulphide conversion fluctuated during process operation (Figure 5.14). After a biofilm disruption occurred, the VSOR rapidly increased to reach a maximum after approximately 24 h, before decreasing to a minimum. This recurring cycle in sulphide oxidation in the LFCR is shown in Figure 5.14 and agrees with the results reported by Mooruth et al. (2013). The oxidation is predominantly attributed to biological oxidation with minimal contribution through abiotic reactions. This was evaluated and confirmed by Mooruth (2013)

The maximum VSOR could not be sustained and sharply decreased over time to a minimum. The decline in sulphide oxidation was consistent with the development of the surface biofilm. The relationship between oxygen mass transport across the FSB and sulphide oxidation rate, described by Mooruth (2013), support these findings. In the current investigation, the results clearly show that once the FSB is established there is a decrease in the VSOR due to the impedance of oxygen mass transport. During the study, an increase in the maximum VSOR (2 L: 0.047 to 0.116 mmol/L.h; 8 L: 0.079 to 0.125 mmol/L.h) was observed as HRT was decreased from 4 to 2 days, respectively. A correlation could be drawn between the VSRR and the maximum VSOR as a function of HRT. As the HRT was incrementally decreased from 4 to 2 days an increase in VSRR was accompanied by an increase in the VSOR. In the hybrid LFCR, the VSRR effectively represents the sulphide loading rate that provides the source sulphide for oxidation. Therefore, as the VSRR increased, the increased availability of sulphide facilitated higher oxidation rates being achieved.

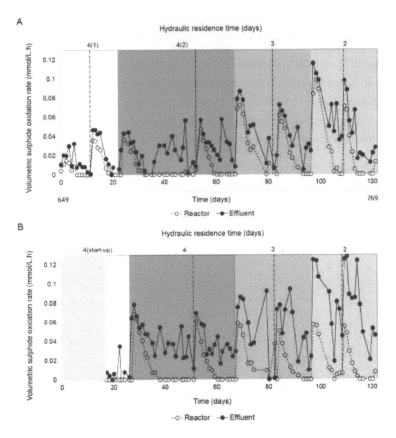

Figure 5.14: Effect of HRT on sulphide oxidation showing the conversion and volumetric sulphide oxidation rate over time in the A) 2 L and B) 8 L lactate-fed reactors. Biofilm disruption and harvest events are indicated by vertical dotted and solid lines, respectively. A change in HRT is indicated by the transition in shading intensity and was accompanied by a biofilm harvest.

Partial sulphide oxidation in the hybrid LFCR is highly dependent on the sulphide to oxygen ratio. Since the open system is operated semi-passively, the amount of oxygen introduced into the system is not controlled. Therefore, management of the FSB is critical to regulate the flux of oxygen at the air-liquid interface to favour partial sulphide oxidation. This is regulated by allowing sufficient sulphide to accumulate in the bulk volume before biofilm disruption to ensure a sulphide to oxygen ratio >2:1, which favours partial oxidation to elemental sulphur. In the current study, disruption of the biofilm more regularly, during experimental run 4(1), had an adverse effect on biological sulphate reduction providing minimal time for the SRB to recover from the perturbation. In contrast, an extended period of operation, during experimental run 4(2), the biofilm became oxygen limiting to a point at which low VSOR was sustained, accumulating high concentration of sulphide within the bulk volume, resulting in reduced sulphide removal and biofilm recovery to sulphide-S ratio.

Although the maximum VSOR corresponded to a sulphide conversion of approximately 80% (Figure 5.14), across the range of HRT tested in both reactors the total sulphide conversion over the duration of each experimental run ranged between 44 to 48% and 49 to 64% in the 2 L and 8 L LFCR, respectively (Table 5.3). Overall the 8 L LFCR performed better than the 2 L LFCR in terms of sulphate reduction and sulphide conversion.

Table 5.3: Summary of overall process performance of the 2 L and 8 L lactate-fed reactors as a function of HRT.

HRT (days)	Volumetric sulphate loading rate (mmol/L.h)	Volumetric sulphate reduction rate (mmol/L.h)	Sulphate conversion (%)	Volumetric sulphide oxidation rate (mmol/L.h)		Sulphide conversion (%)
				Maximum	Average	
2 L LFCR						
4(1)	0.108	0.048	44	0.047	0.022±0.02	40
4(2)	0.108	0.068	63	0.057	0.038±0.02	44
3	0.145	0.083	58	0.087	0.051±0.03	48
2	0.217	0.113	52	0.116	0.073±0.03	47
8 L LFCR						
4	0.108	0.070	65	0.079	0.033±0.02	49
3	0.145	0.095	66	0.092	0.053±0.03	64
2	0.217	0.132	61	0.128	0.080±0.03	52

5.4.2.5 Effect of harvesting regime and hydraulic residence time on biofilm recovery

The biofilm harvested and the total load of sulphide available for oxidation, in grams of sulphur (sulphide-S), are shown in Figure 14. In experimental run 4(1), the shorter operation between biofilm disruption and harvest resulted in higher biofilm recovery to sulphide-S ratio compared to the extended operation during experimental run 4(2), owing to the latter becoming oxygen limiting over time, affecting sulphide oxidation in the biofilm.

A decreasing trend in biomass recovery was observed across both the 2 L (4.7 to 2.1 g FSB) and 8 L (8.2 to 4.4 g FSB) reactor as the HRT was decreased from 4 to 2 days. The biomass harvested within the 8 L reactor was substantially greater than that recovered from the 2 L reactor as expected based on the larger reactor size and surface area. However, the 8 L LFCR was less efficient in biofilm recovery to sulphide-S ratio in comparison to the 2 L reactor.

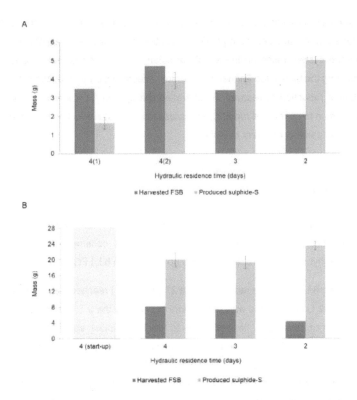

Figure 5.15: Biofilm harvested as a function of hydraulic residence time recovered from the A) 2 L and B) 8 L lactate-fed reactors. The produced sulphide-S represents the cumulative sulphide generated estimated based on sulphate reduction (as per weight sulphur (grams)) over the duration of the experimental run. The error bars represent the standard deviation.

The results re-emphasised the importance of regulating FSB disruption and harvesting within the hybrid LFCR. Previous studies by Mooruth (2013) and Molwantwa (2008) that operated the LFCR as a sulphide oxidising unit, treating a sulphide-rich effluent, suggested that the ideal harvesting period of the biofilm could be as frequent as every 24 h. The results from the current investigation are consistent with these findings, where the maximum VSOR is reached approximately 24 h after biofilm disruption. Ideally the biofilm disruption should occur just before VSOR begins to decline. However, at this point the biofilm is still underdeveloped and has a sticky consistency that make harvesting not practical.

Mooruth (2013) determined that harvesting the biofilm after every 2-3 residence times was optimal in achieving effective sulphur recovery. Results from experimental run 4(1) support these findings, as disrupting the biofilm after every 3 residence times facilitated a higher biofilm recovery to sulphide-S ratio (Figure 5.15). However, in the current hybrid LFCR configuration, combining sulphate reduction and sulphide oxidation within a single reactor, sulphate reduction was adversely affected during biofilm disruption. Since the partial sulphide oxidation within the hybrid system relies on the supply of sulphide via biological sulphate reduction activity, harvesting the biofilm too frequently adversely affected sulphate reduction, limiting the time for sulphate reduction to recover. By operating the system for a longer period before biofilm disruption, in experimental run 4(2), an increase in sulphate reduction was observed at the expense of sulphide-S recovery to biofilm.

5.5 Conclusion

In conclusion, the impact of altered aspect ratio and reactor volume showed minimal effect on system performance over the HRT range considered, thus confirming the stability and robustness of the hybrid LFCR process on scale up from a 2 to 8 L LFCR.

The start-up of the 8 L LFCR was rapid, after just 27 days it had reached stable performance, equivalent to the 2 L LFCR. This was much more rapid than the 2 L reactor. This marked decrease in start-up time can be attributed to the high initial sulphate and sulphide concentration present within the 8 L inoculum. This may have induced a selective pressure, favouring the growth and activity of SRB. In addition, the high sulphide concentration facilitated the rapid formation of the FSB and establishment of anoxic conditions in the bulk volume, providing a suitable environment for SRB.

Stoichiometric analysis of sulphate reduction within the reactor revealed that complete oxidation was favoured within the study. This confirms the establishment of an active acetate SRB community within the LFCR. Few studies have reported complete oxidation of lactate within sulphidogenic reactors. Most studies have reported the limitations of lactate as a carbon source due to the accumulation or low consumption of acetate. In addition, the results showed a discrepancy within the stoichiometric analysis based on lactate metabolism which suggested that an additional carbon source may have been present within the feed. This is further investigated in Chapter 6.

The 2 L LFCR had been operated continuously for 649 day prior to the commencement of the investigation. Although this system did not recover the sulphate reduction performance achieved in Section 4.4 subsequent to operation at an HRT of 12 h, the ability of the system to maintain stable performance over the extended period of operation demonstrated process stability and resilience.

The results confirm that HRT plays a critical role in the overall microbial activity. At a shorter HRT, the system did not allow adequate reaction time to reach high conversion efficiency; however, it promoted high VSRRs. The data suggest that the maximum VSRR may be further increased with a lower HRT as also demonstrated in Section 4.4, albeit at the cost of conversion efficiency. Therefore, based on the compromise between rate and conversion, the choice of operating HRT should consider the desired water quality and treatment rate.

Disruption and harvesting of the biofilm more frequently allowed for higher biofilm recovery. However, an extended period of operation between FSB disruption and harvesting promoted higher sulphate conversion. Based on these results, it is proposed that achieving high biological sulphate reduction should be favoured over regulating biofilm disruption to enhance sulphur recovery and an additional LFCR employed downstream to remove excess sulphide. Since most of the sulphate reduction would occur in the primary reactor, the biofilm in the second reactor could potentially be harvested more frequently. Thus, high biological sulphate reduction and partial sulphide oxidation may be achieved without compromising either process. The operation of a dual hybrid LFCR system incorporating an additional operational unit is explored in Chapters 8 and 9.

Chapter 6
Effect of electron donor

6.1 Introduction

Previous studies evaluating biological sulphate reduction, using a lactate-based feed, observed that the competition between lactate fermenters and lactate oxidising SRB is a key factor impacting overall BSR performance (Oyekola et al., 2012; Bertolino et al., 2012). Lactate is known as an effective carbon source for high sulphate reduction activity and is the substrate of choice for sulphate reduction in terms of energy and biomass yields as well as for its selection of diverse SRB (Celis et al., 2013). Lactate was selected for this study to ensure successful demonstration of the integrated process. Its use as sole carbon source provided a base case for comparison with Oyekola et al. (2010)'s study that used lactate as a sole carbon source for sulphate reduction within a CSTR. Although recognised as an ideal substrate for SRB, its application at large scale is limited by its high cost and availability. Lactate predominantly undergoes incomplete oxidation during sulphate reduction and is inefficient in terms of carbon utilisation per mole of sulphate reduced, resulting in the accumulation of acetate (Celis et al., 2013). The low acetate consumption and acetate accumulation associated with incomplete oxidation of the substrate is a major constraint of lactate-based BSR processes. Based on these parameters, lactate is not favourable as a viable carbon source for industrial application, particularly in wastewater treatment.

It has been suggested that the development and selection of a microbial consortium containing SRB that are complete oxidisers, readily consuming acetate, may overcome a major drawback of lactate as an electron donor and carbon source, making more attractive, efficient and feasible. In wastewater treatment, effective treatment requires reduction of COD to prevent pollution, complete oxidation or two-stage (incomplete-complete) oxidation is required to ensure this.

A major challenge for widespread application of BSR at commercial scale is the provision of a cost effective electron donor (Harrison et al., 2014). Its selection is greatly dependent on cost, availability and kinetics (Liamleam & Annachhatre, 2007; Harrison et al. 2014). Research to date has investigated the potential use of a variety of complex carbon-rich waste streams for biological sulphate reduction (Shoeran et al., 2010; Harrison et al., 2014; Sato et al., 2017). These complex substrates require hydrolysis through anaerobic digestion to form easily degradable, low molecular weight monomers metabolised by SRB, such as simple sugars, alcohols and VFAs. In the vast majority of studies, the COD released is made up of a mixture of VFA with different compositions, often dominated by acetate, followed by propionate (Harrison et al., 2014; Chalima et al., 2017).

Acetate is an important intermediate in the anaerobic mineralisation of organic matter (Celis et al., 2013). The production of acetate during biological sulphate reduction is a major drawback of sulphate reducing bioreactors due to the inability of most known SRB to completely oxidise acetate even in the presence of excess sulphate. The low acetate oxidation efficiency results in low sulphide and alkalinity production and contributes to the high residual COD in the effluent (Liamleam & Annachharte, 2007; Celis et al., 2013). Though acetate is well documented as a suitable carbon source for SRB and is a major component released during the breakdown of complex substrates, few studies have evaluated the use of acetate as a sole carbon source for the application of BSR. The major limitations associated with its use include the selection of SRB capable of complete acetate oxidation and their slow growth kinetics. As a result, the complete oxidisers in an acetate-fed sulphidogenic bioreactor inoculated with a mixed microbial consortium are often outcompeted by faster growing microorganisms or require a long start-up (Celis et al., 2013). Therefore, the composition and activity of the inoculum plays a key role in the successful development of an effective acetate-fed sulphate reducing bioreactor. Furthermore, the presence of alternative carbon sources may impact the process.

In the current chapter the potential application of acetate as an alternative carbon source to lactate for sulphate reduction in the hybrid LFCR was evaluated. This involved a comparative analysis based on process performance in terms of biological sulphate reduction and partial sulphide oxidation.

Key objectives addressed in this chapter include:

1) Investigate the effect of yeast extract on process performance and carbon balance
2) Evaluate the use of acetate as an alternative electron donor in the hybrid LFCR to lactate
3) Assess the effect of hydraulic residence time as a function of electron donor on process performance

This chapter is arranged as follows: the initial start-up of the 2 L acetate-fed LFCR is discussed, followed by the effect of yeast extract and hydraulic residence time process performance. A comparative assessment of acetate as an alternative carbon source to lactate is presented.

6.2 Acetate as an alternative carbon source

6.2.1 Experimental approach

A 2 L LFCR, described in Section 3.1.1, was inoculated with an active acetate-acclimatised inoculum. This SRB inoculum was derived from the original culture (Section 3.1.1), adapted to acetate and maintained in batch culture on acetate as a sole carbon source, with regular sub-culturing at 50% (v/v) with synthetic medium supplemented with 0.92 g/L sodium acetate to provide a feed acetate concentration of 11.2 mmol/L. The active batch culture was expanded to 2 L through sub-culturing to facilitate a full reactor volume inoculation. In addition, sufficient dissolved sulphide concentration (approximately 6.0 mmol/L) could accumulate to promote rapid development of the FSB following inoculation of the LFCR. Adaptation of the inoculum occurred over a period of approximately 3 months. Upon start-up, the reactor was operated at a 4 day HRT and 30°C. Process performance was monitored regularly with sampling (Section 3.1.3) and analysis (Section 3.2) as used earlier.

6.2.2 Results and discussion

The 2 L acetate-fed LFCR was inoculated and operated alongside the lactate-fed LFCRs with the initial 212 days of operation of the 2 L acetate-fed reactor shown in Figure 6.1. Similar to the start-up of the 8 L lactate-fed reactor (Section 5.4.2), accumulation of sulphate within the batch culture, due to sub culturing with a concentrated medium, was higher than the 1 g/L (10.4 mmol/L) feed sulphate concentration. The first 64 days of continuous operation showed an initial period of instability in sulphate concentration due to the acclimatisation, adaptation and colonisation of the inoculum. The continuous operation over time facilitated the gradual washout of excess residual sulphate, resulting in stable volumetric sulphate loading across the system.

Over the first 24 hours of operation the initial sulphide concentration rapidly decreased with the simultaneous development of a structurally sound FSB at the surface. The subsequent increase in sulphide concentration during the start-up phase was a strong indication of sulphate reduction activity. The sulphide concentration increased gradually over time, conforming to a similar profile previously observed within the 2 L and 8 L lactate-fed reactors (Section 5.4.2). After the first biofilm collapse on day 30, the increase in sulphide concentration became more pronounced as a result of SRB activity.

The sulphate concentration increased dramatically on day 65 and 97, respectively, following biofilm harvest and disruption. This resulted in a marked increase in sulphate, reaching an equivalent concentration to that of the feed. Thereafter, the sulphate concentration within the bulk volume deceased as the sulphide concentration increased over time. This was distinctly different from the lactate-fed systems, where minimal to slight increase in sulphate concentration occurred after a biofilm disruption or harvest. The 2 L acetate-fed LFCR reached sulphate concentrations of 5.62 and 4.73 mmol/L, accounting for 46 and 55% sulphate conversion on days 97 and 125, respectively. Interestingly, after biofilm harvest on day 125 the sulphate concentration was less affected by biofilm disruption. The increased biomass concentration and colonisation of carbon microfibers within the reactor enhanced system robustness and SRB activity. This was evident by rapid formation of the FSB at the surface concomitant with the increase in sulphide concentration.

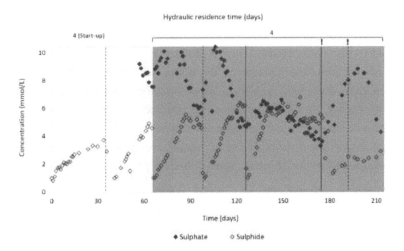

Figure 6.1: Start-up of the 2 L acetate-fed LFCR showing the average dissolved sulphide and residual sulphate concentration measured in the reactor samples (FM, FB, BM, and BB). The initial 65 days represents the start-up and acclimatisation phase (grey shaded area) at a 4 day HRT. Biofilm disruption and harvest events are indicated by vertical dotted and solid lines, respectively.

The system was operated for an extended period (50 days) without any disruption to the biofilm until day 175, with the intention to increase sulphate reduction by maintaining anoxic conditions favourable for SRB activity in the bulk volume over a longer time period. In addition, due to the absence of biofilm disruption, a high sulphide concentration of approximately 5 mmol/L could be maintained. The sulphate concentration gradually decreased to 3.7 mmol/L, equivalent to 64% sulphate conversion. Similar to the lactate study presented in Section 5.4, by increasing operational time between biofilm disruption events, the sulphate reduction performance was increased. In comparison, at a 4 day HRT, the 2 L lactate-fed reactor achieved a sulphate conversion of 63% (Section 5.4.2). Hence, similar system performance could be obtained using acetate or lactate as an alternative carbon source. These findings reiterate the importance of regulating operational time between biofilm collapse events. Without biofilm disruption, the system was able to maintain anaerobic conditions and high sulphide concentration over an extended period, providing selective pressure favouring SRB activity. The inhibitory effects of sulphide on microorganisms are well documented, showing that SRB have a higher tolerance to sulphide than methanogenic and fermentative microorganisms have (Greben et al., 2004; Moosa & Harrison, 2006).

Following the extended period of operation, the biofilm was harvested on day 175. During the early stages of subsequent FSB development, a blockage in the effluent pipe caused the reactor volume to increase. This compromised the structural integrity of the biofilm with a large portion of the biofilm lost to the effluent after the blockage was released and excess volume decanted. Since a structurally sound biofilm was not present at the surface, unimpeded oxygen transfer into the bulk volume affected SRB activity, resulting in a rapid increase in sulphate concentration. The formation of the biofilm was adversely affected after the disturbance, which resulted in a poorly developed biofilm due to the low sulphide concentration (1.3 mmol/L). On day 192, the biofilm prematurely collapsed and settled onto the harvesting screen. The sulphate concentration continued to increase reaching 8.8 mmol/L by day 198, after which system performance began to recover over time. The residual sulphate concentration decreased to 4.28 mmol/L by day 213, corresponding to a sulphate conversion of 59%.

From these results it became evident that the 2 L acetate-fed LFCR was more sensitive to operational perturbation, introduced through biofilm disruption and the influx of oxygen, than the lactate-fed LFCRs. The 2 L acetate-fed system required an extended period to recover sulphate reduction performance after every biofilm disruption event. Previous studies have described the acetate-utilising SRB (complete oxidisers) as more sensitive to environmental conditions than lactate-utilising SRB (incomplete oxidisers) (Celis et al., 2013, Rubio-Rincon et al, 2017). This has often been attributed to their slow growth rate and susceptibility to stress conditions, including exposure to oxygen (Celis et al., 2013, Rubio-Rincon et al., 2017).

6.3 Effect of yeast extract and hydraulic residence time

6.3.1 Experimental approach

In Sections 4.5.2 and 5.4.2, discrepancies in the VFA carbon balance was identified based on stoichiometric analysis. Using an average propionate concentration observed in both 2 L and 8 L LFCRs, the amount of lactate consumed for fermentation was that required for 20% sulphate removal at 1 g/L under incomplete oxidation. Together with the excess accumulation of acetate beyond the theoretical maximum generated based on feed lactate concentration, these results suggested that an unaccounted for carbon source was present in the synthetic feed.

In the current work, a modified Postgate B (MPB) medium was applied; this has been historically used for cultivating SRB and evaluating biological sulphate reduction (Postgate, 1984). MPB is a nutrient rich medium and contains 1 g/L yeast extract (YE), a source of nitrogen, vitamins and growth stimulating compounds for microbial cultivation (Zarei et al. 2016). It contains approximately 4 – 13% (w/w) carbon. While MPB is preferred for SRB cultivation, it is not ideal in kinetic and stoichiometric studies to assess BSR as a function of carbon sources and electron donors, including acetate, lactate and ethanol. In large-scale passive or semi-passive biological treatment systems, YE supplementation is not feasible economically. The impact of YE on sulphate reduction is typically ignored in bioreactor studies employing MPB as synthetic feed. Such YE supplementation may result in an inaccurate measure of specific carbon utilisation and associated sulphate reduction performance where not considered. Saez-Navarrete et al. (2009) evaluated the effect of varying YE concentrations, as the sole source of carbon, on sulphate reducing activity in batch experiments using a pure culture of *Desulfobacterium autotrophicum*. The cultures were started at an initial sulphate concentration of 2.2 g/L and YE concentrations ranging from 0 – 2 g/L. A linear increase in VSRR with increasing YE was observed, with a VSRR greater than 8.3 mg/L.h at YE concentrations of 0.5 and 1 g/L. These findings highlight the potential for YE, as an additional carbon source, on biological sulphate reduction.

In most bioreactor studies of sulphate reduction, carbon source utilisation is not evaluated on a compound basis; instead, measurement of chemical oxygen demand (COD) is used, measuring the oxidisable organic material present in solution (Abba et al., 2017). While this provides a satisfactory measure of nutrient removal, it provides no details on specific carbon utilisation and the metabolic pathways driving the biochemical processes.

In Section 5.4.2, analysis of VFA concentration profiles across the 2 L and 8 L lactate-fed reactors identified inconsistencies in the carbon balance. This was largely attributed to the high acetate concentration detected above the theoretical production via lactate metabolism. Based on these findings it was postulated that an alternative carbon source present within the feed was likely metabolised to acetate. Studies by Chen & Dong (2005) reported the preference for YE as a source of carbon by a proteolytic microorganism *Proteiniphilum acetatigenes* producing acetate. In the current study, the potential of YE as an alternative carbon source in the MPB feed was considered; YE was reduced from 1 to 0.4 g/L to evaluate its impact on the performance of the hybrid LFCR. More importantly, the study aimed to resolve the discrepancies observed within the VFA carbon balance. Before studying YE concentration and evaluating the effect of HRT, all three LFCR systems were placed on a 5 day HRT for consistency across the study through re-establishing high sulphate conversion in both the 2 L and 8 L lactate-fed reactors after the extended operation at a 2 day HRT (Section 5.4). The operation at longer residence times are favourable for promoting SRB activity and biomass retention. Under a long HRT, carbon limiting conditions favour the growth of SRB over fermenters due to the former's high affinity for scavenging the carbon source at low substrate concentrations. The accumulation of sulphide concentration at long residence times also plays an important role, inhibiting non-SRB species at high concentrations (Oyekola et al., 2010). Once stable performance was achieved at a 5 day HRT, the YE concentration was decreased to 0.4 g/L and its effect on process performance was evaluated. The reactors were operated until stable performance was re-established and thereafter subjected to a range of HRTs from 5 to 2 days. Due to the long recovery period (76 days) required to re-establish stable performance within the 2 L acetate-fed reactor, applying a strict biofilm disruption and harvest regime was not practical. Instead, the study evaluated sulphate reduction performance as a function of HRT across all three reactor systems, with minimal disturbance to the FSB.

6.3.2 Results and discussion

6.3.2.1 Effect of hydraulic residence time on process performance

At the start of the current investigation at a 5 day HRT, a biofilm harvest was performed to initiate the study at day 0. An increase in residual sulphate concentration was observed across all reactors (Figure 6.2). In the 2 L and 8 L lactate-fed reactors, the sulphate concentration increased over 18 days reaching concentrations of 8.2 and 7.3 mmol/L, respectively. Shortly after, the sulphate concentration rapidly decreased to an average concentration of 1.8 mmol/L and 1.4 mmol/L by day 52, equivalent to approximately 83 and 87% sulphate conversion, respectively. Prior to the commencement of the current investigation, the 2 L and 8 L lactate-fed reactors, had been continuously operated for a total of 903 and 254 days, respectively. Despite having been operated continuously at a 2 day HRT, both 2 L and 8 L lactate-fed systems were able to recover high sulphate conversion on a 5 day HRT. The increase in residual sulphate concentration after the FSB harvest was most pronounced within the 2 L acetate-fed LFCR, reaching an equivalent concentration to that of the feed (10.4 mmol/L) on day 30. An extended lag phase, with minimal change in sulphate concentration occurred up until day 59. This was followed by a decrease until day 100, achieving a residual sulphate concentration and corresponding conversion of 3.4 mmol/L and 67%, respectively. From the time of biofilm harvest, the acetate-fed LFCR required nearly double the time (100 days) to recover stable performance compared to the lactate-fed reactors (52 days).

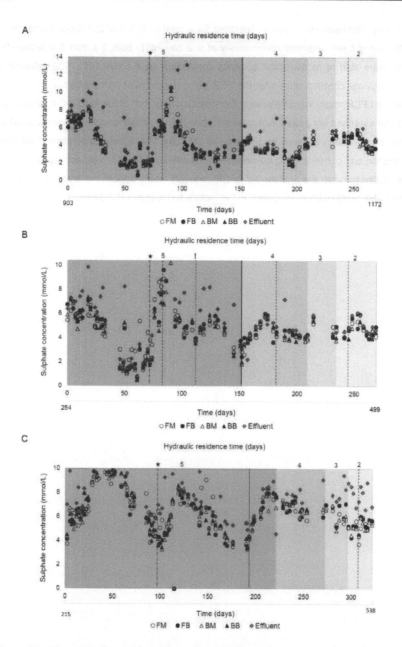

Figure 6.2: Effect of HRT on residual sulphate concentration measured over time in the A) 2 L lactate-fed, B) 8 L lactate-fed and C) 2 L acetate-fed reactors. Biofilm disruption and harvest events are indicated by vertical dotted and solid lines, respectively. A change in HRT is indicated by the transition in shading intensity. The adjustment in YE concentration is represented by (★) while premature disruption by (!). The second x-axis indicates the time of operation relative to start-up.

The impact of biofilm harvest on sulphate concentration was attributed to oxygen inhibition on the SRB population, where biofilm disruption resulted in an influx of oxygen into the bulk volume. A study by Rubio-Rincon (2017) reported partial inhibition of sulphate reducing activity after exposure to oxygen (2.7 mg O_2/L). They performed oxygen tests on batch cultures fed with lactate, acetate and propionate. Although similar inhibition activities (approximately 53%) across carbon sources were found, the longest inactivation time (lag phase) was observed in the acetate culture (1.75 h), while no lag phase was observed in the lactate-fed culture. Rubio-Rincon (2017) concluded that the inhibitory and toxic effects on SRB are dependent on the carbon source and demonstrated that activity can be recovered on restoring anaerobic conditions.

Similarly, in the current study, after biofilm disruption within the hybrid LFCR, the SRBs were adversely affected by the increased transfer of oxygen into the bulk volume. Although both lactate- and acetate-fed reactors were affected, the acetate-fed reactor was most sensitive to the perturbation. After the biofilm was regenerated and anaerobic conditions restored, sulphate conversion recovered. The extent to which SRB overcome oxygen inhibition has been linked to their ability to metabolise organic matter. Thermodynamically, according to Gibbs free energy, SRB generates twice the amount of energy during incomplete oxidation of lactate (-160.3 kJ/mol S) compared to its complete oxidation (-84 kJ/mol S) (Rubio-Rincon, 2017). Ramel et al. (2015) determined the molar growth yields of *Desulfovibro vulgaris* Hildenborough (incomplete oxidiser) on lactate under oxic conditions and concluded that some of the energy gained from lactate oxidation was directed toward cell protection against oxidative conditions and associated repair, rather than biosynthesis. Incomplete oxidisers have a lower doubling time (3-10 h). The ability of lactate to select for a diverse SRB community and generate high biomass concentration is reported to confer greater versatility and tolerance to environmental stress. In contrast, complete oxidisers are less versatile and generally express a higher sensitivity, due to high doubling time (16-20 h) (Celis et al., 2013). SRB are able to endure short term exposure to oxic conditions through different strategies including: 1) reduction of oxygen to water by membrane-bound terminal oxidases (Ramel et al. 2015), 2) adherence to biofilms where the formation of gradients reduce the exposure to oxygen (Rubio-Rincon, 2017) as well as 3) the potential symbiosis with aerobic microorganisms (sulphur oxidising bacteria) (Bade et al., 2000).

After re-establishing stable sulphate conversion at a 5 day HRT, the YE concentration was changed from 1 g/L to 0.4 g/L on day 72 and 98 in the lactate- and acetate-fed reactors, respectively. As seen in Figure 6.2, the sulphate concentration increased across all three reactors on perturbation. In the 2 L lactate- and acetate-fed reactors, the sulphate concentration increased to a maximum of 8.0 and 8.2 mmol/L, respectively. Thereafter, the residual sulphate concentration decreased over time, achieving the prior sulphate conversion. Comparable to the response to biofilm harvest at the beginning of the investigation (day 0), the 2 L acetate-fed reactor exhibited a greater sensitivity to the change in YE concentration, requiring a longer recovery period (76 days) than the 2 L lactate-fed reactor (40 days).

In the 8 L lactate-fed reactor, a blockage of the effluent port on day 112, due to the build-up of elemental sulphur and biofilm fragments, caused an increase in reactor volume. After releasing the blockage, the excess volume was discharged. During this period, the structural integrity of the FSB was compromised resulting in a premature disruption and subsequent collapse of the FSB onto the harvesting screen. Consequently, the sulphate concentration increased to 5.5 mmol/L on day 124, before gradually decreasing to 2.4 mmol/L on day 152. As a result, a total of 80 days was required to re-establish stable performance, achieving a sulphate conversion equivalent to 77%.

Since the biofilm was not disrupted when the YE concentration was adjusted, the marked increase in sulphate concentration suggests that decreasing YE directly affected metabolic activity of the SRB population. Although the decrease in YE concentration influenced sulphate reduction, all three reactors were able to recover a similar performance observed before the adjustment.

Once steady state performance was re-established after the adjustment in YE concentration, the FSB was harvested in all three reactors. This occurred on day 191 in the acetate-fed reactor and on day 152 in both lactate-fed reactors. The HRT was then incrementally decreased over time, with biofilm harvest only conducted again toward the end of the study at a 2 day HRT. Due to the effect of biofilm harvest on the 2 L acetate-fed reactor, the change in HRT to a 4 day was delayed until the sulphate conversion began to recover. The lactate-fed reactors experienced fluctuations in sulphate concentration, as the HRT was decreased, on average ranging between 3.0-5.0 mmol/L and 3.7-5.8 mmol/L in the 2 L and 8 L lactate-fed reactors, respectively. The 2 L acetate-fed reactor exhibited a gradual decreasing trend in sulphate concentration, particularly during operation at 5 and 4 day HRT. This was attributed to the system recovering from the biofilm harvest.

The dissolved sulphide concentration profiles corresponded well with the sulphate data (Figure 6.2). Similar to the results obtained in Sections 4.4 and 5.4, disruption to the biofilm resulted in a sharp decrease in dissolved sulphide over 24 h. As the biofilm regenerated at the air-liquid interface it formed a barrier to oxygen penetration in the bulk volume, resulting in the accumulation of sulphide over time

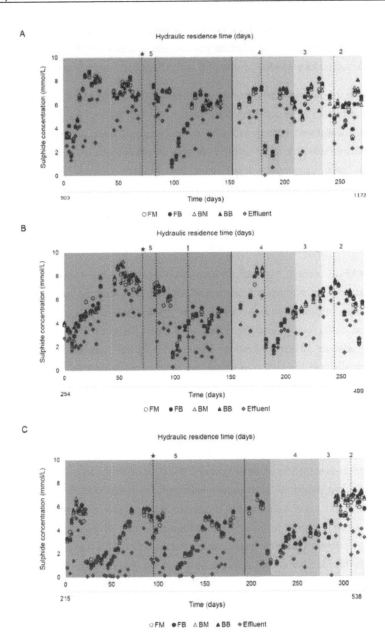

Figure 6.3: Dissolved sulphide concentration profile showing reactor samples (FM, FB, BM, BB) and effluent in the A) 2 L lactate-fed LFCR, B) 8 L lactate-fed and C) 2 L acetate-fed reactors. A change in HRT is indicated by the transition in shading intensity. The adjustment in YE concentration is represented by (★) while premature disruption by (!). The second x-axis indicates the time of operation relative to start-up.

For the purpose of the current investigation, the frequency of biofilm disruption and harvesting was reduced. Consequently, high sulphide concentrations accumulated within the reactors and were maintained for extended periods until biofilm disruption was performed.

Consistent with the sulphate data, minimal variation in sulphide concentration was observed between the different reactor sampling ports, namely FM, FB, BM and BB (Figure 6.3). This indicated that a well-mixed, homogenous environment was maintained within the bulk volume of the reactors with little to no stratification of aqueous chemical species. The effluent sulphide concentration was consistently lower than the reactor samples. Higher sulphide concentrations were measured in the lactate-fed reactors compared to the acetate-fed reactor, owing to improved sulphate reduction. On average, the maximum sulphide concentration, after biofilm disruption, ranged between 6.0 and 9.0 mmol/L in the lactate-fed reactors and between 5.0 and 7.0 mmol/L in the acetate-fed reactor.

In the hybrid LFCR process, the sulphate conversion determines the available sulphide for sulphide oxidation. By comparing experimental sulphide data and the expected sulphide (calculated from sulphate reduced) as an area graph shown in Figure 6.4, a visual representation of the relative sulphide conversion can be observed. The graph displays a comparative analysis of the theoretical amount of sulphide generated based on the corresponding sulphate conversion as well as the average dissolved sulphide in the reactor and final effluent over time. The amount of sulphide conversion is therefore the difference between the expected and effluent sulphide concentration.

From the results sulphide is converted in the bulk volume and final effluent based on the expected sulphide. The most pronounced fraction of sulphide conversions occurred after a biofilm disruption event within all three reactors. Based on cumulative sulphide conversion calculated over the period for each experimental run, the acetate-fed reactor had a high conversion than that observed within the lactate-fed reactors. The conversion is dependent on the amount of expected sulphide which fluctuated based on the degree of sulphate conversion, therefore it should be noted that the lactate-fed reactors had a higher expected sulphide. The results also concur with previous findings that illustrate that the sulphide conversion becomes restricted overtime as the biofilm becomes oxygen limiting. The cumulative sulphide conversion obtained in this study is summarised in Table 6.4 and will be discussed together with the sulphide oxidation kinetics in Section 6.3.2.4.

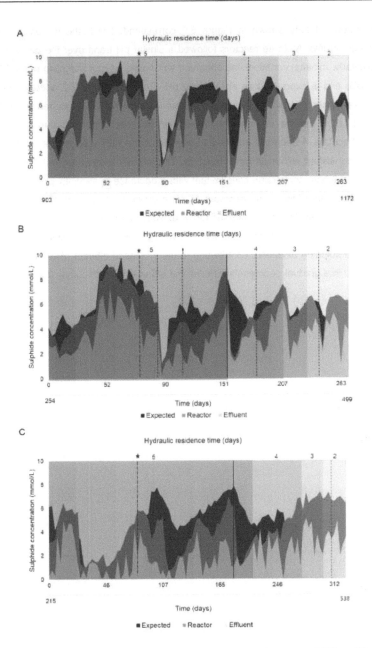

Figure 6.4: Sulphide dynamics showing expected sulphide generated, reactor sulphide and final treated effluent sulphide concentration over time A) 2 L lactate-fed, B) 8 L lactate-fed and C) 2 L acetate-fed reactors. A change in HRT is indicated by the transition in shading intensity. The adjustment in YE concentration is represented by (★) while premature disruption by (!). The second x-axis indicates the time of operation relative to start-up.

The measured pH data shown in Figure 6.5 corresponded with the dissolved sulphide concentration profile. All three reactors followed a similar pH trend over the duration of the study. Similarly, minimal variation was observed among the reactor sampling ports (FM, FB, BM and BB) and is expressed as an average pH. On average the 2 L acetate-fed LFCR maintained a higher pH than the lactate-fed systems. The higher pH measured within the 2 L acetate-fed LFCR can be attributed to the higher alkalinity generated through complete oxidation of acetate. A decrease in pH was observed as the HRT was decreased from 4 to 2 days, was consistent within all three reactors. This corresponded well with the observed reduction in sulphate conversion concomitant with a decrease in alkalinity production as the HRT decreased. In addition, the decrease in pH coincided with the increase in measured acetate concentration. The 8 L lactate-fed LFCR exhibited a higher effluent pH in comparison to the reactor samples. The greater surface area within the 8 L LFCR configuration facilitated higher partial sulphide oxidation, contributing to the increased effluent pH. The decrease in HRT resulted in a gradual decrease in reactor and effluent pH, particularly from a 4 to 2 day HRT.

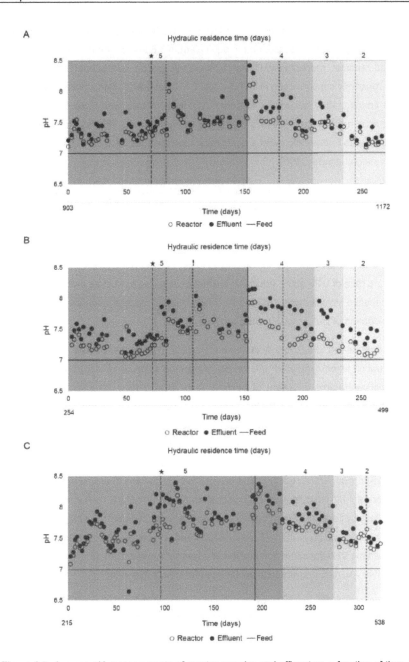

Figure 6.5: Average pH measurements of reactor samples and effluent as a function of time A) 2 L lactate-fed, B) 8 L lactate-fed and C) 2 L acetate-fed reactors. A change in HRT is indicated by the transition in shading intensity. The adjustment in YE concentration is represented by (★) while premature disruption by (!). The second x-axis indicates the time of operation relative to start-up.

In the previous chapter, toward the end of the HRT study, the operation at a 2 day HRT resulted in a high propionate concentration equivalent to 3.8 and 3.6 mmol/L in the 2 L and 8 L lactate-fed reactors, respectively. In the current investigation, during operation at a 5 day HRT, low propionate concentration in both 2 L and 8 L lactate-fed systems was observed. The propionate concentration decreased to approximately 2 mmol/L within the 2 L lactate-fed LFCR and was not detected within the 8 L lactate-fed reactor. The results revealed the decrease in lactate fermentation with a concomitant increase in biological sulphate reduction during operation at a 5 day HRT.

The adjustment in YE concentration, on day 72, had implications on the overall VFA profiles, in all three reactors. There was a marked decrease in acetate concentration from 15.7 to 9.0 mmol/L and 16.3 to 7.0 mmol/L, in the 2 L and 8 L lactate-fed LFCR, respectively. This was accompanied by a sulphate concentration of 3 and 2.9 mmol/L with a sulphate conversion of 71 and 72% in the 2 L and 8 L lactate-fed LFCRs, respectively. This corresponded to a stoichiometric ratio of 0.78 and 1 (mol/mol) sulphate converted to acetate produced and was higher than the 0.5 stoichiometric ratio required for incomplete lactate oxidation, indicating that complete oxidation of lactate to CO_2 occurred.

In the 2 L acetate-fed LFCR, the acetate concentration (Figure 6.6 C) was not affected by biofilm disruption when compared with the corresponding sulphate concentration (Figure 6.2 C). The relatively stable utilisation of acetate, even after biofilm disruption, suggests the presence of an active non-SRB microbial population capable of acetate metabolism within the reactor. Alternatively, it may represent rapid complete re-oxidation of sulphide to sulphate while sulphate reduction still occurred. Following the adjustment in YE, a decrease in acetate concentration, from 8.5 to 1.5 mmol/L, was observed and was noted to be consistent with that of the lactate-supplemented reactors. The acetate-fed reactor was operated at a feed acetate concentration of 11.2 mmol/L. The findings confirm the impact of YE on the VFA profile and the accumulation of acetate. Since the excess acetate was not detected in the feed it is likely that the carbon source present in YE was metabolised to acetate. Although the decrease in YE influenced the VFA concentration profiles and sulphate conversion, all three reactors were able to recover a similar performance achieved before the adjustment.

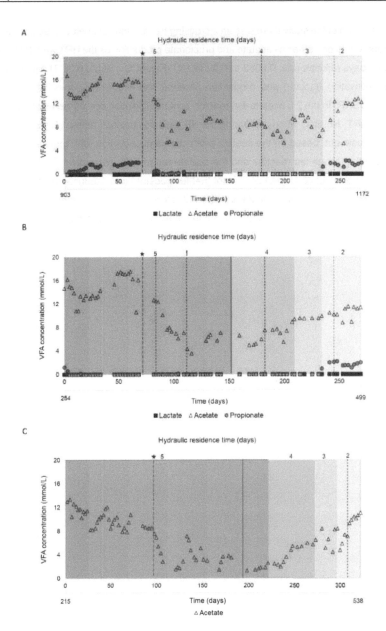

Figure 6.6: Effect of HRT on the volatile fatty acid concentration profile measured over time in the A) 2 L lactate-fed, B) 8 L lactate-fed and C) 2 L acetate-fed reactors. A change in HRT is indicated by the transition in shading intensity. The adjustment in YE concentration is represented by (★) while premature disruption by (!). The second x-axis indicates the time of operation relative to start-up.

Over the duration of the study there was an increasing trend in the volumetric rates associated with lactate utilisation as well as acetate and propionate production as the HRT was decreased from 5 to 4 days (dilution rate: 0.0083 to 0.0208 1/h) (Figure 6.7; Table 6.2). The increase in substrate utilisation (lactate) and production (acetate and propionate) coincided with the increase in VSRR. In the 2 L acetate-fed reactor, acetate utilisation decreased when exposed to a 2 day HRT. Notably, when comparing the YE feed composition before (1 g/L) and after (0.4 g/L) adjustment, based on the acetate production and utilisation rate, there was a clear shift in both the lactate-fed reactors and the acetate-fed reactor, respectively. A decrease in acetate within the lactate-fed reactor meant that less acetate was produced as a product of incomplete lactate oxidation. In the 2 L acetate-fed reactor where acetate served as the main carbon source, the decrease in acetate meant that a higher portion of acetate was utilised.

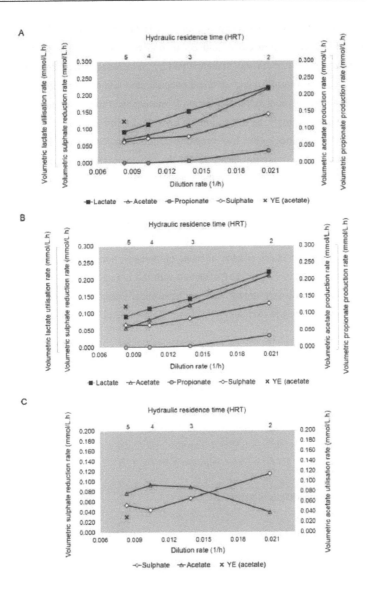

Figure 6.7: Effect of HRT on the volumetric sulphate reduction, substrate (lactate and acetate) utilisation and production (acetate and propionate) rates in the A) 2 L lactate-fed, B) 8 L lactate-fed and C) 2 L acetate-fed reactors. YE represents the molar ratio before the adjustment in concentration from 1 to 0.4 g/L.

Across the range of dilution rates applied, complete lactate utilisation was maintained throughout the study (Figure 6.7 A and B). The trend in acetate utilisation, as HRT decreased, coincided with that in sulphate reduction. At low dilution rate, corresponding to high sulphate conversion (Figure 6.10 A and B), low concentrations of lactate and acetate indicates complete oxidation was favoured (Figure 6.6 A and B). In contrast, at high volumetric loadings a decrease in sulphate conversion was accompanied by an increase in acetate concentration. At 2 day HRT, a sudden increase in propionate concentration within both lactate-fed reactors was observed, an indication of increased lactate fermentation. The propionate concentration reached 1.85 and 1.90 mmol/L by the end of the investigation on day 269 in the 2 L and 8 L lactate-fed reactors, respectively. The propionate concentrations were lower than that reported in Section 5.4.2. These results agree with the findings of Oyekola et al. (2012), where operation at higher dilution rates favoured the growth of fermentative microorganisms causing a shift in the active microbial population and dominant lactate metabolic pathway.

6.3.2.2 Stoichiometric dependency on hydraulic residence time

As in Section 5.4.2.2, the stoichiometric analysis on electron donor utilisation and sulphate reduction is summarised in Table 6.2. Consistent with the previous study, the experimental ratios L:A, L:S, A:S coincided largely with the theoretical values (Table 6.1) of complete (Reaction 6.1) and incomplete oxidation (Reaction 6.2) toward sulphate reduction.

Table 6.1: Theoretical stoichiometric ratios of the metabolic reactions associated with the respective electron donors.

Reaction No.	Chemical reaction	Theoretical stoichiometric ratio		
		L:A	L:S	A:S
6.1	$2\ Lactate^- + 3\ SO_4^{2-} \rightarrow 6\ HCO_3^- + 3\ HS^- + H^+$	-	0.67	-
6.2	$2\ Lactate^- + SO_4^{2+} \rightarrow HS^- + 2\ Acetate^- + 2\ HCO_3^- + H^+$	1.0	2.0	2.0
6.3	$3\ Lactate \rightarrow acetate + 2\ propionate + HCO_3^- + H^+$	3.0	-	-
6.4	$Acetate^- + SO_4^{2-} \rightarrow HS^- + 2\ HCO_3^-$	-	-	1[a]

[a] acetate utilised per mol sulphate reduced

Table 6.2: Effect of HRT on the molar ratio of lactate utilised to moles of acetate and propionate produced involved in biological sulphate reduction, using lactate as the sole carbon-source and electron donor. Average values of experimental stoichiometric ratios are compared with the theoretical ratios (Table 6.1). The carbon balance of total moles VFA accounted compared with the total amount of lactate fed is also presented.

HRT (days)	Volumetric rates (mmol/L.h)				Stoichiometric ratios			Carbon balance
	Lactate utilisation rate	Acetate production rate	Propionate production rate	Sulphate reduction rate	Total moles lactate used/mole acetate produced (L:A)	Total moles lactate used/mole sulphate reduced (L:S)	Total moles acetate produced/ mole sulphate reduced (A:S)	Total C moles out/ Total C mole lactate fed [b] (Effluent:Influent)
2 L lactate-fed								
5	0.091	0.124	0.011	0.072	0.7	1.3	1.7	1.0
5 [a]	0.091	0.069	0.001	0.062	1.3	1.5	1.1	0.5
4	0.114	0.081	0.000	0.072	1.4	1.6	1.1	0.5
3	0.152	0.111	0.006	0.078	1.4	2.0	1.4	0.5
2	0.222	0.219	0.035	0.144	1.0	1.5	1.5	0.8
8 L lactate-fed								
5	0.091	0.121	0.001	0.075	0.8	1.2	1.6	0.9
5 [a]	0.091	0.058	0.000	0.067	1.6	1.4	0.9	0.4
4	0.114	0.081	0.000	0.066	1.4	1.7	1.2	0.5
3	0.143	0.125	0.003	0.085	1.1	1.7	1.5	0.6
2	0.222	0.211	0.034	0.129	1.0	1.7	1.6	0.8

[a] Yeast extract concentration adjustment to 0.4 g.L
[b] Carbon balance of the total mol C measured (residual lactate, acetate and propionate) to total amount of mol C lactate fed

The change in YE affected the stoichiometric profiles (Figure 6.8 and 6.9) in all three reactors. In the lactate-fed reactor the most noticeable impact can be observed by the low L:A ratio which suggests that there was more acetate measured in the system than expected if the feed lactate concentration was metabolised via incomplete oxidation (Reaction 6.2) or fermentation (Reaction 6.3). This together with the observed decrease in the carbon balance (effluent: influent carbon ratio) confirmed that YE had influenced the overall carbon load within the hybrid LFCR. These findings were consistent with the concentration profiles seen in Figure 6.6.

When excluding the contribution of fermentation (SR) based on residual propionate concentration, the stoichiometric ratios were relatively similar to that obtained when the contribution of fermentation was included (F) (Figure 6.8). However, the experimental ratios shifted with HRT. The biggest deviation between F and SR ratios was observed among the estimated L:S and A:S experimental ratios (Figure 6.8 B) at a 2 day HRT, which coincided with the observed increase in propionate concentration (Figure 6.6). In the 2 L lactate-fed reactor, the L:S ratio including the contribution of fermentation (F) agreed with the theoretical value of incomplete oxidation while the L:S ratio excluding the contribution of fermentation (SR) affiliated more closely with complete lactate oxidation.

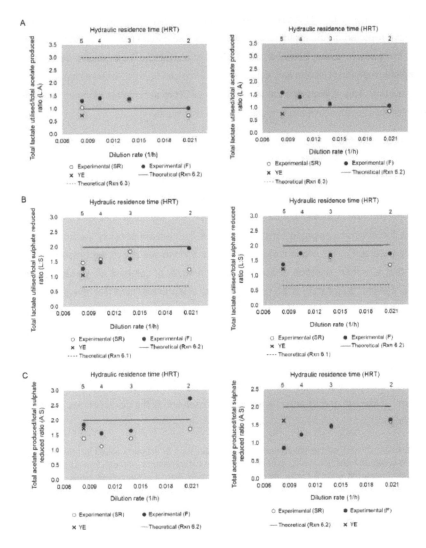

Figure 6.8: Effect of HRT on biological sulphate reduction stoichiometry in the 2 L (left) and 8 L (right) reactors, showing the A) Total moles of lactate utilised per mole total acetate produced (L:A), B) total moles of lactate utilised per total moles sulphate reduced (L:S), C) moles of acetate produced per total moles of sulphate reduced (A:S). Experimental ratio with (F) and without (SR) the contribution of fermentation, calculated stoichiometrically (Rxn 6.3) based on residual propionate concentration. The horizontal solid (Rxn 9.2) and dotted (9.1 and 9.3) lines represent the theoretical ratio for the respective reactions. YE (x) represents the molar ratio before the adjustment in concentration from 1 – 0.4 g/L.

These results demonstrate the importance of accounting for the contribution of fermentation to accurately represent the sulphate reduction stoichiometry in the lactate-fed systems. The low A:S ratios estimated in the 2 L and 8 L reactors indicates that both complete and incomplete oxidation of lactate toward sulphate reduction occurred within the system. Several studies have reported on the limitation of lactate as a sole carbon source and the inability to establish complete oxidation (Oyekola, 2008; Celis et al., 2013). These results demonstrate that complete oxidation of lactate as a sole carbon source can be established within a sulphate reducing system. This is likely due to the ability of the LFCR to accumulate biomass facilitating the establishment of an active acetate-utilising SRB community.

In the 2 L acetate-fed reactor, high acetate utilisation was favoured at the longer HRT. Similarly, as observed in the lactate-fed reactors, the decrease in YE affected the A:S ratio and carbon balance (Table 6.3; Figure 6.9). As the HRT was decreased to 2 days, the A:S ratio decreased coinciding with the decrease in acetate utilisation and reduction in sulphate conversion. The low A:S ratio before the YE adjustment, indicates that less acetate was utilised for the observed sulphate conversion than is theoretically possible via Reaction 6.4. After the adjustment in YE concentration the A:S ratio increases above the theoretical. This revealed that a higher utilisation of acetate to the amount of sulphate converted occurred within the system, suggesting a portion of acetate metabolised was not directed toward sulphate reduction. It is expected that the microbial community will comprise of other non-SRB microorganism that are capable of metabolising acetate and will compete with the SRB community. These findings concur with the results obtained in the lactate-fed reactors where the adjustment in YE reduced the impact on the overall VFA balance within the system. By varying the YE available, data was collected on its use, particularly by a fermentative mechanism. This data and its analysis allowed for resolution of the discrepancies in the VFA mass balance.

Table 6.3: Effect of HRT on the volumetric rate and molar ratio of acetate utilised to moles sulphate reduced via sulphate reduction, using acetate as the sole carbon-source and electron donor. Average values of experimental stoichiometric ratios are compared with the theoretical ratios (Table 6.1; Reaction 6.4). The carbon balance of total moles acetate measured to total moles of acetate fed into the system is also shown

HRT (days)	Volumetric rates (mmol/L.h)		Stoichiometry	Carbon balance
	Acetate utilisation rate	Sulphate reduction rate	Total Moles acetate used/ mole sulphate reduced (A:S)	Moles of acetate out/ moles of acetate in [b] (Effluent: Influent)
2 L acetate-fed LFCR				
5	0.030	0.055	0.5	0.7
5 [a]	0.077	0.053	1.4	0.3
4	0.093	0.044	2.1	0.3
3	0.089	0.067	1.3	0.5
2	0.039	0.115	0.3	0.8

[a] Yeast extract adjustment from 1 – 0.4 g/L
[b] Carbon balance of the total mol acetate measured to the total amount of mol acetate fed

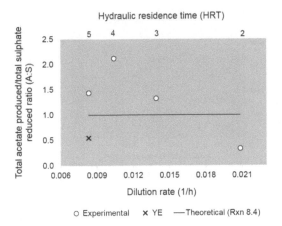

Figure 6.9: Effect of HRT on biological sulphate reduction stoichiometry via Reaction 6.4 in the 2 L acetate-fed reactor, showing the total moles of acetate utilised per mole sulphate reduced (A:S). The horizontal solid line represents the theoretical A:S ration based on Reaction 6.4. YE (x) represents the experimental ratio before the adjustment in concentration from 1 – 0.4 g/L.

6.3.2.3 Biological sulphate reduction kinetics

The sulphate reduction kinetics are presented in Figure 6.10. At the end of the experiments reported in Section 5.4.2, while operated at a 2 day HRT, the 2 L and 8 L lactate-fed reactors achieved a VSRR of 0.113 and 0.132 mmol/L.h which was equivalent to a sulphate conversion of 52 and 61% conversion, respectively. In the current investigation, operation at a 5 day HRT resulted in an increase in sulphate conversion to 83 and 87%. However, due to the decrease in dilution rate, the VSRR decreased to 0.072 and 0.075 mmol/L.h, respectively. These results demonstrated the resilience of the hybrid process to recover high sulphate conversion when operated at a low dilution rate (5 day HRT). In contrast, the 2 L acetate-fed LFCR was less effective compared to its lactate counterpart and achieved a VSRR of 0.055 mmol/L.h with a sulphate conversion of 63% at a 5 day HRT.

The reduction in YE concentration impacted the residual VFA concentration, which was highlighted by the decrease in acetate concentration across all reactors (Figure 6.6) and a sharp increase in residual sulphate concentration (Figure 6.2), showing that sulphate reduction activity was affected negatively. All reactors recovered a performance similar to that achieved before the reduction in YE. This was equivalent to a maximum VSRR of 0.062 and 0.067 mmol/L.h with a conversion of 71 and 77% in the 2 L and 8 L lactate-fed LFCR, respectively. In the 2 L acetate-fed LFCR, although the recovery period was nearly double the time, a VSRR of 0.053 mmol/L.h with a corresponding conversion of 62% was achieved.

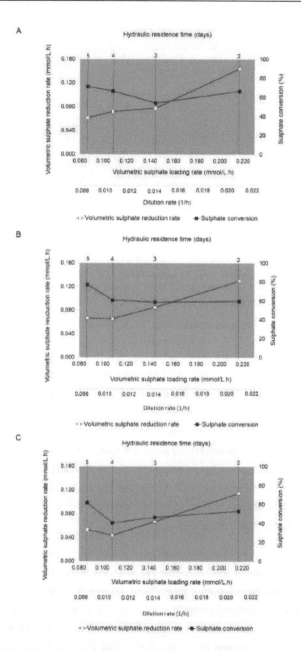

Figure 6.10: Steady state kinetics of sulphate reduction performance as a function of hydraulic residence time, showing the volumetric sulphate reduction rate and sulphate conversion in the A) 2 L lactate-fed, B) 8 L lactate-fed and C) 2 L acetate-fed reactors.

As the reactors were subjected to decreasing HRT conditions, an overall increase in VSRR was accompanied by a decrease in sulphate conversion. In the lactate-fed LFCRs there was an increase in VSRR (2 L: 0.062 to 0.078 mmol/L.h; 8 L: 0.067 to 0.085 mmol/L.h) concomitant with a decrease in sulphate conversion (2 L: 71 to 54%; 8 L: 77 to 59%) from 5 to 3 day HRT. The VSRR increased once operated at a 2 day HRT achieving 0.144 and 0.129 mmol/L.h with a corresponding sulphate conversion of 66 and 60%. The 8 L lactate-fed reactor performed slightly better than the 2 L lactate-fed reactor. Although the sulphate reduction performance in the 2 L acetate-fed reactor was constantly lower than the lactate-fed reactors, a similar response to changing HRT conditions was observed in all three reactors. In the 2 L acetate-fed LFCR, a decrease in HRT from 5 to 3 days resulted in an increase in VSRR from 0.053 to 0.067 mmol/L.h accompanied by a decrease in sulphate conversion from 62 to 46%. A further decrease in HRT to 2 days, resulted in a marked increase in VSRR to 0.115 mmol/L.h with a corresponding sulphate conversion of 53%.

6.3.2.4 Biological sulphide oxidation kinetics

The sulphide oxidation kinetics in the hybrid LFCR is driven by the volumetric sulphide concentration available for oxidation which is dictated by the VSRR. In addition, sulphide oxidation is controlled through management of biofilm disruption, which regulates the influx of oxygen and the accumulation of dissolved sulphide in the bulk volume. Results from Section 4.5.2 and Section 5.4.2 demonstrated the importance of regulating biofilm disruption to ensure maximum sulphide removal and sulphur recovery. A key parameter is to ensure a sulphide to oxygen ratio >2, to favour partial sulphide oxidation.

Due to the restricted biofilm disruption regime that was adopted in the current investigation, where biofilm disruption was kept to a minimum, the evaluation of the sulphide oxidation performance was limited. The VSOR profiles over the duration of the study is presented in Figure 6.11. The effluent samples were highly variable in comparison to the average VSOR among the reactor samples. This was due to the unpredictable discharge rate at the outlet port which affected the final effluent sulphide concentration and the corresponding VSOR. The higher VSOR in the final effluent was expected, since the reactor volume is directed toward the surface and out the effluent port, which represents the aerobic zone where sulphide oxidation takes place. The VSOR profile, based on the reactor samples, exhibited a similar trend described in Section 5.4. After a biofilm disruption or harvest event, the VSOR rapidly increased to a maximum. The high VSOR could not be maintained owing to decreasing oxygen transfer through the FSB as it thickened and subsequently decreased over time to an absolute minimum. The decrease in VSOR was consistent with the formation of the FSB as it matured at the surface and became oxygen limiting.

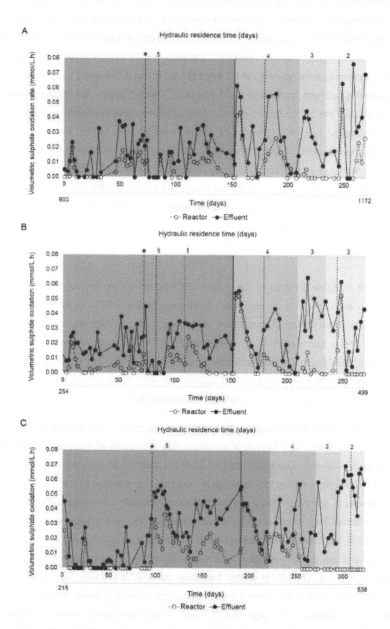

Figure 6.11: Effect of HRT showing the volumetric sulphide oxidation rate over time in the A) 2 L lactate-fed, B) 8 L lactate-fed and C) 2 L acetate-fed reactor. A change in HRT is indicated by the transition in shading intensity. The adjustment in YE concentration is represented by (★) while premature disruption by (!). The second x-axis indicates the time of operation relative to start-up.

Since a consistent biofilm disruption regime was not performed between each HRT condition, the effect on the sulphide oxidation could not be evaluated. However, there was a distinct increase in maximum VSOR in the 2 L and 8 L lactate-fed reactors, when the HRT was changed from 5 to 4 days, from 0.02 to 0.044 mmol/L.h and 0.024 to 0.052 mmol/L.h in the 2 L and 8 L lactate-fed reactors, respectively. This agreed with the results in Section 5.4.2.4, where the VSOR increased as HRT was decreased. Interestingly, the 2 L acetate-fed reactor exhibited a higher VSOR when operated at a 5 day HRT compared to that of the lactate-fed reactors, after the YE was reduced, reaching a maximum of 0.044 mmol/L.h. The higher VSOR observed in the 2 L acetate-fed reactor was consistent with the cumulative sulphide conversion, over the duration of each experimental run, calculated based on the difference between the expected sulphide and final effluent concentration (Table 6.4).

Table 6.4: Overall process performance comparing reactor geometry and use of different electron donor as a function of HRT. Results summarise the effect of operational conditions on sulphate reduction performance and sulphide conversion.

HRT (days)	Volumetric sulphate loading rate (mmol/L.h)	Volumetric sulphate reduction rate (mmol/L.h)	Sulphate conversion (%)	Volumetric sulphide oxidation rate (mmol/L.h) Maximum	Volumetric sulphide oxidation rate (mmol/L.h) Average	Sulphide conversion (%) [b]
			2 L Lactate-fed			
5	0.087	0.072	83	0.038	0.014±0.01	19
5 [a]	0.087	0.062	71	0.035	0.018±0.01	33
4	0.108	0.072	66	0.062	0.027±0.02	39
3	0.145	0.078	54	0.045	0.029±0.01	28
2	0.217	0.144	66	0.077	0.034±0.03	29
			8 L Lactate-fed			
5	0.087	0.075	87	0.045	0.017±0.01	32
5 [a]	0.087	0.067	77	0.045	0.023±0.01	34
4	0.108	0.066	61	0.056	0.026±0.02	41
3	0.145	0.085	59	0.050	0.030±0.02	39
2	0.217	0.129	60	0.065	0.034±0.02	26
			2 L acetate-fed			
5	0.087	0.055	63	0.056	0.019±0.02	52
5 [a]	0.087	0.053	62	0.056	0.024±0.01	71
4	0.108	0.044	40	0.057	0.025±0.02	50
3	0.145	0.067	46	0.059	0.030±0.02	34
2	0.217	0.115	53	0.070	0.059±0.01	49

[a] Yeast extract concentration adjustment 1 – 0.4 g/l
[b] Cumulative conversion based on expected and final effluent sulphide concentration

6.4 Conclusion

Despite the difference in size and aspect ratio, the 2 L and 8 L LFCR maintained similar performance throughout the study. Both reactors, when subjected to an operational perturbation, showed a similar response in process performance. The ability of the lactate-fed LFCRs to recover high sulphate conversion once operated at a 5 day HRT, demonstrated the robustness and resilience of the hybrid LFCR process. The results confirm that high sulphate reduction can be restored when operated at low dilution rates that favour the activity of SRB.

The adjustment in YE resolved the discrepancies observed within the VFA carbon balance, particularly the excess acetate concentration. Though this had major implications on the VFA concentration profiles and initially on the sulphate conversion, all three reactors were able to recovery similar performance before the change. However, even at a 60% reduction, a concentration of 0.4 g/L YE may still have a considerable impact on the overall performance. This was evident by its contribution on the measured acetate in all three reactors. The form of carbon present within YE and to what degree it contributes towards the observed sulphate reduction performance was not determined.

Throughout the duration of the experiment, acetate as a carbon source was less effective than lactate. In addition, the 2 L acetate-fed LFCR was more susceptible to operational perturbations introduced through biofilm disruption or the adjustment in YE. On average, it took double the time to recover performance within the 2 L acetate-fed reactor compared to the lactate-fed reactors. The system was able to maintain a relatively stable sulphate conversion when exposed to decreasing HRT. The ability to maintain sulphate reduction without substantial loss in performance when exposed to higher dilution rates can be attributed to the LFCRs ability to retain high biomass concentrations.

Stoichiometry analysis over the duration of the study revealed that complete oxidation in the lactate-fed reactors was present over the range of HRT evaluated. This was consistent with the results obtained in Section 5.4.2.2. In the 2 L acetate-fed reactor the decrease in YE resulted in a shift in the stoichiometric ratio towards the theoretical stoichiometry for complete oxidation. In addition, the adjustment in YE concentration resulted in a decrease in the ratio of effluent to influent carbon (effluent: influent), measured as VFAs, which was more representative of the feed lactate and acetate concentration. Although 0.4 g/L of YE may still have an impact on the availability of carbon, to maintain consistency with subsequent studies and ensure availability of growth factors, the YE remained unchanged and was accounted for in data analysis.

The effect of HRT on biological sulphate reduction revealed that system performance must be governed as a compromise between VSRR and sulphate conversion. As HRT is decreased and VSRR increases, the sulphate conversion decreases. Therefore, depending on the application and the desired quality of water, the choice of HRT should consider both performance values to facilitate select process performance. These results are supported by the findings obtained in Section 5.4.

The results from the current investigation as well as Section 4.5 and 5.4 reiterate the importance of the microbial community in dictating the overall performance of the hybrid LFCR. This was particularly highlighted by the shift in metabolic degradation of lactate observed under changing HRT conditions that affected the overall sulphate reduction. In addition, the consistent acetate utilisation in the 2 L acetate-fed reactor, higher than the theoretical for the observed sulphate conversion, strongly indicated the presence of an active non-SRB population. It has been reported that the major contribution to poor sulphate conversion in acetate-fed sulphidogenic bioreactors is linked to the microbial ecology.

Based on these findings, the complex microbial community dynamics present within the hybrid LFCR has been suggested to be an important factor, as it drives overall process performance. The change in process performance as a function of operational parameters is directly associated with the shift in microbial community structure that govern the biochemical reactions, explored in Chapter 7.

Chapter 7

Microbial community dynamics

7.1 Introduction

During the experimental studies conducted in Section 4.5 and 5.4, it became increasingly apparent that microbial community dynamics plays a critical role in the overall performance of the hybrid LFCR process. This was highlighted by the change in reactor performance, reaction stoichiometry and VFA profile as the HRT was changed. Based on these findings, an investigation into microbial ecology using next generation 16S rRNA amplicon sequencing was initiated concurrently with the experiments conducted in Section 6.3. The intention was to evaluate the microbial community dynamics within the hybrid LFCR process as a function of HRT across reactor design and electron donor.

The study of microbial community response to changing conditions has become an increasingly important parameter in evaluating bioprocesses, particularly in wastewater treatment (Vasquez et al., 2018). Sulphate reducing bacteria (SRB) and sulphur oxidising bacteria (SOB) have been studied extensively in different natural environments for their vital role in both the sulphur and carbon cycles and, more recently, the involvement of SOB in the nitrogen cycle (Zhang et al., 2017; Nielsen et al., 2018; Vasquez et al., 2018). However, little information is available regarding the occurrence of highly diverse SRB and SOB communities within bioreactor systems (Zhang et al., 2017). In the hybrid LFCR discrete microenvironments facilitate the development of distinctly separate SRB and SOB microbial communities that drive the biochemical reactions of the process. Traditionally biological wastewater treatment has been operated by black box approach with minimal consideration of the microbial biocatalysts that drive the chemical reactions. Advances in molecular biology have revolutionised the ability to study microbial communities with regards to composition and diversity as well as understanding the roles of microorganisms within a given environment (Logue et al., 2015). The latest advancements in NGS technologies in the field of metagenomics represent a powerful approach for resolving complex microbial communities (Cui et al., 2017).

In current work, the hybrid LFCR represents a unique environment where both biological sulphate reduction and sulphide oxidation occurs within close proximity in a single operational unit. It represents one of a few environments where the coexistence of SRB and SOB communities and synergistic dynamics can be studied. In the following study, the microbial community dynamics within the hybrid LFCR process was evaluated as a function of HRT by applying next generation 16S rRNA sequencing.

The main objectives addressed in this study are outlined as follows:

- Determine the microbial community structure and relative abundance of key species within the hybrid LFCR;
- Evaluate the changes in microbial community structure in response to hydraulic residence time and its effects on system performance;
- Assess the effect of reactor design and scale-up on the microbial ecology; and
- Evaluate the effect of different electron donors on microbial ecology.

7.2 Experimental approach

Microbial community analysis using next generation 16S rRNA amplicon sequencing was performed on a total of 8 samples collected from the 2 L and 8 L lactate-fed reactors as well as the 2 L acetate-fed reactor. Samples were taken from each reactor, based on discrete microenvironment phases to facilitate the establishment of different microbial communities within the hybrid LFCR (Figure 7.1). This included the biomass attached to the carbon microfibers (CF), planktonic phase (PV) made up of free floating microorganisms within the bulk volume region associated with the carbon microfibers, the planktonic phase (PS) just below the air-liquid interface and the FSB. Samples were taken from the HRT study reported in Section 6.3 at stable performance at a 5 and 2 day HRT to evaluate the effects of HRT, reactor scale and geometry and electron donor on the microbial community. A total of 24 samples were sequenced. The samples were processed as described in Section 3.6.

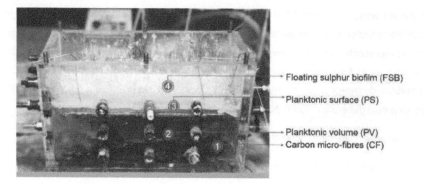

Figure 7.1: Sampling layout of the selected samples (CF, PV, PS and FSB) taken after operation at 5 and 2 day HRT for 16S rRNA sequencing. The sampling was performed on the 2 L and 8 L lactate-fed reactors as well as the 2 L acetate-fed reactor. The discrete niche environments that were sampled across the reactor are shown.

7.3 Results and Discussion

7.3.1 Microbial α and β diversity analysis

High-throughput 16S rRNA sequencing resulted in a total of 365 operational taxonomic units (OTUs), identified across 24 samples collected from four locations in three reactors at two time points. The number of OTUs detected in each sample ranged from a minimum of 82 to a maximum of 173. The bacterial richness (Chao1 estimator) and diversity (Shannon and Simpson indices) estimates were calculated for each sample and are summarised in Table B1 (Appendix B). The Goods coverage estimate ranged from 99.8 to 99.9% indicating that the sequencing depth for all samples was sufficient while the average read count was approximately 19,000. The diversity indices (Shannon and Simpson) as well as the species richness index (Chao 1) indicated that all samples were relatively diverse with similar evenness across all samples. Previous studies, based upon the use of traditional approaches (which included DGGE and clone libraries), underestimated microbial diversity (Zhang et al., 2017). However, the application of high-throughput sequencing based on the 16S rRNA gene overcomes many of the limitations associated with traditional techniques, providing a deeper sequencing depth and almost complete coverage (Good's coverage values greater than 99%) of the microbial communities as obtained in the current study. This provides a more comprehensive analysis with regards to the diversity and community composition.

Beta-diversity analysis was evaluated using the weighted Unifrac algorithm and represented by principal coordinates analysis (PCoA). UniFrac is a distance metric which measures the phylogenetic distance between sets of taxa in a phylogenetic tree. The method determines whether microbial communities are significantly different and the contribution of different factors that result in variation (Lozupone and Knight, 2005). The PCoA technique provides an informative visualisation of the data structure and variation between samples based on community composition and relative abundance (Figure 7.2). In order to determine the effect of HRT on the microbial community structure, weighted PCoA analysis was performed at a 5 day and 2 day HRT. The results at a 5 day HRT revealed a clear separation of the microbial communities taken from different locations within the three operating hybrid LFCR reactors, with a clear divergence between the lactate-fed reactors and the acetate-fed reactor samples. The FSB samples of all three reactors clustered together and were completely separate from the reactor samples (CF, PV and PS) taken within the BSR zone. It was anticipated that the PS (planktonic at the surface) samples collected just below the air-liquid interface, would represent a transition phase in which the microbial community would exhibit similarities between both reactor volume-derived samples (CF and PV) and the FSB. However, based on PCoA analysis, the PS communities clustered more closely with the bulk volume samples consistently across all three reactors

Figure 7.2: Principal coordinates analysis (PCoA) of the weighted UniFrac distances matrices based on the V3-V4 zone 16S rRNA gene sequences of the representative samples (CF, SRB, SOB, FSB) from the 2 L and 8 L lactate-fed reactors as well as the 2 L acetate-fed reactor at a A) 5 day HRT and B) 2 day HRT.

The response to decreasing the HRT from a 5 day to a 2 day HRT impacted the PCoA plot of the different communities shown in Figure 7.2 B. Though the samples mostly conformed to a similar cluster pattern observed at a 5 day HRT (Figure 7.2 A), the samples were more dispersed. This was particularly more pronounced among the FSB communities of the lactate-fed reactors which diverged towards the communities present within the bulk volume (CF, PV and PS). The results highlight the dynamic behaviour amongst microbial communities across the discrete sampling environments within the hybrid LFCR. The source of electron donor had the most significant influence on the observed variation between the lactate-fed and acetate-fed reactors. Based on the PCoA analysis, the effect of reactor design and geometry between the 2 L and 8 L reactors had minimal impact on the microbial population. This corresponds to the similar reactor performance observed between the two reactors.

7.3.2 Microbial community composition and relative abundance at phylum level

In total, 16 different bacterial phyla represented by 365 OTUs were detected across all 24 samples. The dominant taxa which exhibited a minimum relative abundance ≥1% in at least one sample, was represented by 6 major phyla. On average, this accounted for 95.7% of the total sequences in each sample. The dominant phyla (Figure 7.3) included Proteobacteria, Synergistetes, Bacteroidetes, Firmicutes, Thermotagae and Chlorobi. The taxa belonging to the phyla Thermotogae and Chlorobi were exclusively detected at relative abundance >1% within the 2 L lactate-fed and 2 L acetate-fed reactors, respectively.

The variation in microbial community structure observed at phylum and class level, was consistent with the PCoA analysis. The weighted PCoA analysis is highly dependent on the dominant organisms within a population; the results can be influenced strongly by the relative abundance of a small number of dominant organisms. This was demonstrated by the significant proportion of Alphaproteobacteria detected within the FSB communities (Figure 7.2 A) whose high abundance at a 5 day HRT had a significant impact on the PCoA analysis.

The 2 L and 8 L lactate-fed reactors exhibited a similar microbial community structure at phylum-class level, across all samples. Proteobacteria dominated, with the majority of sequences assigned to classes Deltaproteobacteria and Alphaproteobacteria. At a 5 day HRT, Deltaproteobacteria dominated within the CF, PV and PS communities with an average relative abundance of approximately 35±12% and 39±5% within the 2 L and 8 L lactate-fed reactors, respectively (Figure 7.3 A and B). This was followed by the phyla Bacteroidetes, Synergistetes and Firmicutes. While Firmicutes had a higher proportion within the 2 L lactate-fed reactor, Bacteroidetes was more abundant within the 8 L lactate-fed reactor. The high relative abundance of Deltaproteobacteria within the communities derived from the bulk volume was expected as it represents the largest group of know SRB. This corresponded well with the degree of sulphate reduction performance observed within the reactors. Interestingly, in the 2 L acetate-fed reactor (Figure 7.3 C), the relative abundance of Deltaproteobacteria was significantly lower, ranging between 9 - 17% across the reactor samples CF, PV and PS at a 5 day HRT. Instead, the communities in the 2 L acetate-fed reactor were dominated by the phyla Bacteroidetes (32 - 42%) *and* Synergistetes (17 – 37%).

In the FSB communities, Alphaproteobacteria dominated in the 2 L and 8 L lactate-fed reactors making up ±66% and ±79% of the total microbial community, respectively. Similarly, within the FSB derived from the 2 L acetate-fed reactor, the microbial community was predominantly comprised of Alphaproteobacteria (51%). There was also a significant proportion of Betaproteobacteria (31%) within the 2 L acetate-fed FSB at the 5 day HRT.

The effect of decreasing the operating HRT from 5 to 2 days impacted the microbial structure of the CF, PS and PV communities significantly, resulting in a shift in relative abundance of dominant phyla across all three reactors. In the lactate-fed reactors, the CF communities remained relatively stable after exposure to high dilution rate; however, the planktonic derived communities PV and PS were more dynamic, as expected. Across both lactate-fed reactors, the proportion of Deltaproteobacteria decreased. In the 2 L lactate-fed reactor, this was accompanied by an increase in the relative abundance of the phyla Bacteroidetes and Synergistetes; in the 8 L lactate-fed reactor, the increase in Firmicutes was most pronounced. In the 2 L acetate-fed reactor, the abundance of the dominant phyla Bacteroidetes, and Firmicutes shifted within the planktonic communities. In addition, there was a marked increase in the abundance of Betaproteobacteria and Epsilonproteobacteria, which was only observed in the PS community of the 2 L acetate-fed reactor at a 2 day HRT (Figure 7.3 C).

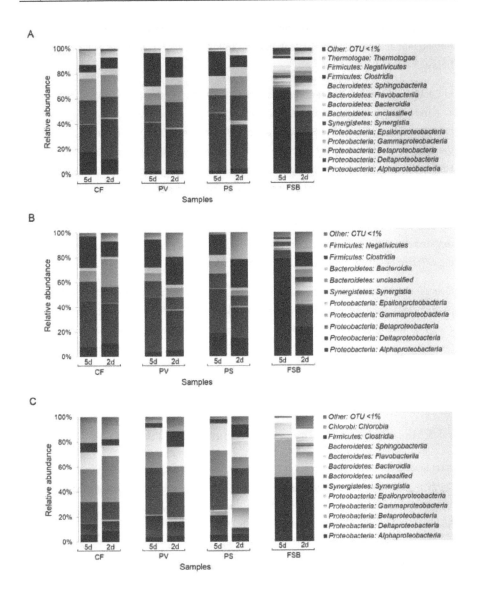

Figure 7.3: Phylum and class level structural changes of the microbial communities (CF, PV, PS, and FSB) as a function of hydraulic residence time at a 5 day and 2 day HRT showing the A) 2 L lactate-fed B) 8 L lactate-fed and C) 2 L acetate-fed reactors.

The FSB was the microbial community which seemed to be the most affected by the change in HRT. Though Proteobacteria remained the most dominant phyla within the FSB there was a shift in the relative abundance of the affiliated classes. Particularly the decrease in the abundance of the class Alphaproteobacteria and increase in the proportion of Betaproteobacteria, Gammaproteobacteria and Epsilonproteobacteria which occurred, between a 5 and 2 day HRT.

In Section 5.4 and 6.3, the performance of the FSB was regulated by the frequency of biofilm disruption and available dissolved sulphide. In the current investigation, the observed microbial composition of the FSB was attributed to the degree of biofilm maturity rather than the effects of decreasing HRT. This is primarily due to the decoupling of the FSB from the HRT in the reactor. Furthermore, at the 5 day HRT, the biofilm was sampled after 68 days of operation while at the 2 day HRT the FSB was sampled after just 12 days i.e. at greatly different maturity of the FSB with different microbial composition observed between communities. A previous study that employed the use of DGGE and clone library sequencing reported the shift in microbial composition of the FSB as the biofilm matured over time through the different developmental stages (Molwantwa, 2008).

Overall, these results revealed the shift in the microbial community structure exhibited at phylum-class level as a function of decreasing HRT from 5 to 2 days. The least affected samples, relative to composition and abundance, were the attached microbial community on the carbon microfibers (CF). In contrast, the FSB samples showed the most variation in microbial community composition between 5 and 2 day HRT. This was consistent with the PCoA results where the most significant divergence was observed between FSB samples at 5 and 2 day HRT.

7.3.3 Microbial classification and distribution at OTU level

OTU level taxonomic classification of 16S rRNA Illumina® MiSeq® sequence data using the QIIME pipeline was performed. QIIME is an open-source bioinformatics pipeline for performing metagenomics analysis on raw DNA sequencing data generated and is described in Section 3.6. The phylogenetic affiliations of the 16S rRNA gene sequences are presented in a heatmap (Figure 7.4). All 55 OTUs identified with a relative abundance >1% within at least one sample is shown. This illustrates the diversity of the microbial communities and the distribution of each classified OTU as a function of HRT and across each reactor. The results highlight the preferential colonisation of specific OTU within the different communities (zones) of the reactor.

The heatmap reveals specific OTUs that were only associated with the FSB while others were detected within the attached and planktonic communities. Overall, there was a distinct shift observed in the relative abundance of the OTUs as a response to change in operating HRT. The OTU taxonomic classification was limited due to the partial sequencing of the 16S rRNA gene. In most cases OTU classification could only be resolved to genus level, while others were limited to phylum and class level identification (Figure 7.4). Alternatively, these OTU may represent novel organisms that have yet to be classified. The heatmap analysis (Figure 7.4) served as an effective tool for illustrating the total microbial community composition and distribution of the OTU representatives detected across the three reactors. However, to better understand the community dynamics within these systems and to assess the effects of HRT, a closer analysis into the relative abundance of the dominant sequences within the different communities was performed.

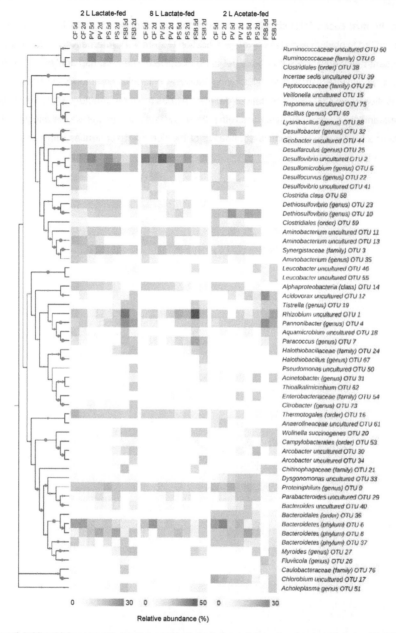

Figure 7.4: Microbial community heatmap analysis based on relative abundance of the classified OTUs (relative abundance ≥1%), showing the effect of HRT within the different microbial communities (CF, PV, PS and FSB) in the 2 L lactate-fed, 8 L lactate-fed, and 2 L acetate-fed reactors. The phylogenetic tree was inferred using the Neighbour-Joining method. Blanch points supported by bootstrap values >50% indicated by size of solid circle. Evolutionary analyses were conducted in MEGA7 (Kumar et al., 2016).

Across all three reactors, there was a diverse complement of SRB genera. While the majority of known SRB are taxonomically conserved within the phylum Deltaproteobacteria, the second largest collection of known SRB are represented within Firmicutes within the Class Clostridia (Tian et al., 2017). The distribution of shared and unique OTUs between the lactate-fed reactors is shown in Figure 7.5. The Venn diagram depicts the change in OTUs and affiliated SRB distribution within the CF and PV communities. The major identified SRB were represented by *Desulfovibrio* (OTU 2) and *Desulfomicrobium* (OTU 5) and were present within both CF and PV communities. As highlighted previously, there was preferential distribution of microorganisms within the different communities.

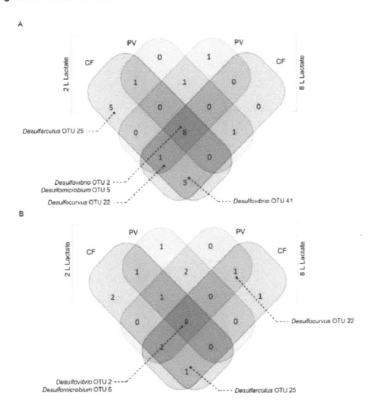

Figure 7.5: Venn diagram showing the shared and unique OTUs >1% between the CF and PV samples comparing the 2 L lactate-fed (blue) and 8 L lactate-fed (purple) reactors at A) 5 day HRT and B) 2 day HRT, the number represents the amount of OTUs while the distribution and location of known SRB are highlighted.

Several OTUs were only present within the attached community on the carbon microfibers (CF) while others were predominantly found within the planktonic phase (PV). The majority of sequences were shared across the lactate-fed reactors. OTUs assigned to genera *Desulfarculus* (OUT 25) and *Desulfovibio* (OUT 41) were only present within the CF communities. The decrease in HRT shifted the distribution of the microbial population between the CF and PV communities. When comparing the 2 L lactate-fed and 2 L acetate-fed reactors, a clear separation of unique OTUs was observed. *Desulfovibrio* (OTU 2) was present within both CF and PV communities (Figure 7.6). Interestingly, *Desulfobacter* was only present within the planktonic phase in the 2 L acetate-fed reactor. Similarly, *Desulfarculus* and *Desulfovibrio* was confined within the CF communities attached onto the carbon microfibers.

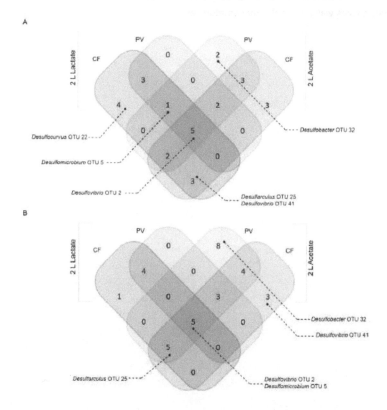

Figure 7.6: Venn diagram showing the shared and unique OTUs between the CF and PV samples comparing the 2 L lactate-fed (blue) and 2 L acetate-fed (red) reactors at A) 5 day HRT and B) 2 day HRT; the number represents the amount of OTUs while the distribution and location of known SRB are highlighted.

The decrease in HRT resulted in a shift in the distribution of shared and unique OTUs across the two reactors. Most notably, there was a marked increase in the quantity of OTUs, particularly within the CF and PV communities derived from the 2 L acetate-fed reactor. Both *Desulfovibrio* (OTU 41) and *Desulfocurvus* (OTU 22) decreased below the 1% limit in relative abundance and thus was not observed at the 2 day HRT (Figure 7.6 B). It is likely that these microorganisms were slow growing and once exposed to high dilution rates were washed-out.

7.3.3.1 Microbial analysis of attached and planktonic communities

Although 16S rRNA sequencing is limited to community structure and composition analysis and does not provide the extended sequence data for determining functional activity and metabolic potential, the presence of well-established phylogenetic groups do provide a strong indication of their possible role within the reactor owing to prior knowledge of dominant metabolic potential of key species. In the current study, 11 classified OTUs were closely related to known SRB belonging to the phylum Deltaproteobacteria (6) and potential SRB genera within the class Clostridia (5) (Figure 7.4). The dominant SRB genera were affiliated to *Desulfovibrio*, *Desulfomicrobium*, *Desulfocurvus*, *Desulfarculus* and *Desulfobacter*. The two most abundant SRB genera detected across all three reactors were *Desulfovibrio* and *Desulfomicrobium* which belong to the group of incomplete organic oxidisers. *Desulfobacter* and *Desulfarculus* are representative of complete organic oxidisers, capable of acetate metabolism. The dominance of *Desulfovibrio* and *Desulfomicrobium* have been reported in several studies to be the most abundant SRB genera in sulphidogenic bioreactors (Hao et al., 2014; Tian et al., 2017; Vasquez et al., 2018).

The phyla Bacterioidetes, Synergistetes and members of Firmicutes are generally associated with fermentative microorganisms that are capable of degrading carbohydrates and volatile fatty acids to acetate and propionate. Some genera have also been characterised as syntrophic acetate oxidising bacteria which are key players within anaerobic bioreactors. These phyla are commonly detected at high abundance within wastewater treatment systems that contain high COD and are responsible for regulating carbon and nitrogen within these environments. Vazques et al. (2018) found that fermentative microorganisms were the most common taxa alongside SRB within sulphidogenic bioreactors. The OTUs classified as *Ruminococcacaea* and *Veilonellea* belonging to the phylum Firmicutes were dominant within the lactate-fed reactors. Member of Firmicutes are largely associated with fermentation; however, a subgroup of SRB taxa is classified within the order Clostridia. Studies have reported sulphate reduction activity by members of the *Ruminococcacaea* and *Peptococcaceae* families within the order Clostridia (Gupta et al., 2018).

In the current investigation, the genus *Veillonella* (OTU 15) belonging to the phyla Firmicutes, was only detected within the lactate-fed reactors. Its presence is consistent with the findings of Oyekola et al. (2012). Upon decreasing the HRT to 2 days, the relative abundance of fermentative microorganisms affiliated with OTUs classified as *Veilonella* (OTU 15),

Synergistetes (OTU 3), and Bacteriodetes (OTUs 6 and 8) increased within both lactate-fed reactors. The increase in *Veilonella* within the PV and PS communities was the most substantial within the 8 L lactate-fed reactor (Figure 7.7 B). This was accompanied by a considerable decrease in the relative abundance of *Desulfovibrio* (OTU 2) as well as the potentially sulphate reducing *Ruminococcaceae* (OTU 0). Interestingly, the second most dominant SRB genus identified as *Desulfomicrobium* (OTU 5) increased as the HRT decreased. The shift in microbial community corresponded well with the performance data, where a decrease in sulphate conversion was observed in the 2 L lactate-fed and 8 L lactate-fed reactors as the HRT was decreased from 5 to 2 days from 71 to 66% and 77 to 60% respectively. This was accompanied by an increase in fermentation evident by the increase in propionate concentration.

The fermentation of lactate to acetate and propionate is well documented in members of the genus *Veillonella* (Stams et al., 2009; Oyekola et al., 2010). Studies have reported the competition between fermenters and SRB in lactate-fed sulphidogenic CSTR bioreactors and the shift in metabolic activity when exposed to change in HRT (Oyekola et al., 2012; Bertolino et al., 2012). Oyekola et al. (2012) determined that lactate oxidisers were characterised by a μ_{max} of 0.2 1/h and K_s value of 0.6 g/L compared with fermenters that exhibited a μ_{max} of 0.3 1/h and a K_s of 3.3 g/L. The study concluded that lactate fermenters outcompeted SRB under conditions of excess lactate and high dilution rate.

The results in the current study highlight that the residence time in the reactor (planktonic cells), rate of substrate loading, as well as the prevailing conditions within the reactors determine which microorganisms preferentially degrade lactate. These results confirm the findings that the operation at high dilution rates, coupled to high lactate and sulphate loading rates, favoured the activity of fermentative microorganisms which resulted in a shift in the microbial community. Alternatively, the operation at low dilution rates favoured the growth and activity of SRB.

The SRB genera *Desulfobacter* (OTU 32) and *Desulfarculus* (OTU 25) were the only known complete oxidisers detected in this study. These were not at the cut off relative abundance limit >1% in the lactate-fed reactors, despite the accumulation of acetate. The accumulation of acetate as a product of incomplete oxidation of the substrate and the inability to select for a complete oxidising, acetate consuming SRB population within sulphidogenic bioreactors has been well reported. This has been defined as one of the major inefficiencies of many sulphate reducing processes that apply the use of ethanol or lactate as an electron donor (Celis et al., 2013).

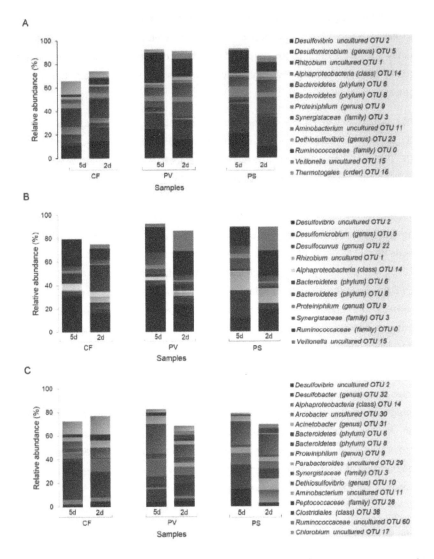

Figure 7.7: Microbial composition at OTU level showing the structural shift in relative abundance of the dominant sequences within the CF, PV and PS communities as a function of decreasing HRT from 5 to 2 days in the A) 2 L lactate-fed B) 8 L lactate-fed and C) 2 L acetate-fed reactors

Deltaproteobacteria, which constituted the major SRB phylum within the lactate-fed reactor, was considerably less prevalent within the 2 L acetate-fed reactor. The only SRB genera within the acetate-fed reactor were represented by *Desulfovibrio* and *Desulfobacter*. Despite the low abundance of the SRB genera, the 2 L acetate-fed reactor achieved sulphate conversion of 62% and 53% at 5 and 2 day HRT, respectively i.e. some 20% less than in the lactate-fed reactors. Interestingly, though *Desulfobacter* was less than 1% within the CF and PS communities, it had a relative abundance of 10% within the PV community at a 5 day HRT. This may suggest preference for the planktonic phase over attached growth on the support matrix. Its absence from the PS community near the surface may indicate its sensitivity to oxygen stress.

Members of Bacteriodetes (38±0.07%) and Synergistetes (27±0.1%) were dominant within the 2 L acetate-fed reactor making up approximately ±65% of the sequences detected within the CF, PV and PS communities. The dominant members of *Synergistetes* were represented by OTUs that could only be classified to the family Synergistacaea, an uncultured *Aminobacterium* and an unidentified *Dethiosulfovibrio* species. A distinguished feature common to all members of the phylum Synergistetes is the capacity to ferment a variety of amino acids as a source of energy. Synergistetes are also involved in the recycling of key nutrients by rapidly digesting proteins of dead microorganisms (Lesnik et al., 2014; Dessi et al., 2018), with the main product of amino acid fermentation being acetate and succinate. The genus *Dethiosulfovibrio*, highly abundant in the 2 L acetate-fed reactor, is capable of sulphur and thiosulphate reduction to sulphide, but unable to use sulphate as an electron acceptor (Magot et al., 1997). This suggests the activity of alternative metabolic pathways in cycling intermediate sulphur species, particularly within the 2 L acetate-fed LFCR.

Under the phylum Bacteroidetes, an OTU classified as *Proteinniphilum*, was consistently detected across all reactors in the CF, PV and PS samples. This strictly anaerobic proteolytic microorganism shows preferred growth on yeast extract (YE) and peptone as a carbon source (Chen & Dong, 2005). Its presence alongside other fermentative microorganisms, predominantly within the phyla Bacteroidetes, Synergistetes and Firmicutes, may account for the degradation of YE and accumulation of acetate above the theoretical level based on feed concentration of electron donor (lactate and acetate).

Measured sulphate reduction was affected by the intermittent biofilm disruption and harvest events. Studies report the isolation of *Desulfovibrio* strains able to survive fluctuating oxygen regimes and resume sulphate reducing activity immediately once anoxic conditions are re-established. The loss in performance may reflect the inability of the SRB community to maintain and recover performance in the presence of O_2, largely dictated by the physiological limitations of complete oxidisers (Rubio-Rincon et al., 2017).

In an oxic-anoxic gradient environment, SRB generally migrate away from high oxygen concentrations. Those found near the anoxi-oxic layer predominantly belong to incomplete oxidising species such as *Desulfovibrio*, *Desulfomicrobium* and *Desulfobulbus* (Sass et al.,

2002). In contrast, the presence of complete oxidising SRB species within oxic zones is rare. A similar observation was seen in the current investigation. Both incomplete oxidising SRB belonging to genera *Desulfomicrobium* and *Desulfovibrio* were detected at high abundance near the air-liquid interface (PS) in the lactate-fed reactors operated at a 5 day HRT. *Desulfomicrobium* exhibited a higher abundance within the PS community than within the CF and PV communities, suggesting preferential proliferation and colonisation within the oxic-anoxic planktonic zone at the air-liquid interface (PS) in response to selective pressure and environmental conditions at the surface i.e. that *Desulfomicrobium* was more adapted to these conditions than microorganisms dominant within the CF and PV communities. Zang et al. (2017) similarly reported the higher oxygen tolerance of *Desulfomicrobium* and its isolation near the oxic/anoxic interface. While several studies report sensitivity and inhibition of SRB when exposed to oxygen stress, some SRB are well adapted to cope with oxygen and can survive temporary exposure as well as reduce oxygen through various mechanisms (Sass et al., 2002). SRB have a natural behavioural response to oxygen, such as migration (chemotaxis) and aggregate formation (Bade et al., 2000; Sass et al., 2002) e.g. *Desulfovirio* species have capability to couple oxygen reduction with proton translocation and energy conservation (Ramel et al., 2015).

Another important strategy for the survival of SRB in an oxic environment is the coexistence with an aerobic microbial population (Bade et al., 2000). In the hybrid LFCR, a synergistic relationship is maintained to sustain the activity of both anaerobic SRB and aerobic SOB communities. While the SRB generate sulphide within the bulk volume creating a narrow redox environment at the air-liquid interface suitable for SOB activity and FSB formation, the SOB consume oxygen as an electron acceptor coupled to the oxidation of the sulphide. This alleviates the inhibitory action of oxygen on the SRB community. Furthermore, as the FSB develops and matures over time, oxygen penetration across the air-liquid interface is impeded, resulting in suitable anoxic conditions for SRB activity within the reactor volume.

Although all SRB conduct dissimilatory sulphate reduction, their substrate utilisation capabilities, kinetics, growth rate and other characteristics such as tolerance to oxygen exposure and other stresses differ, impacting reactor performance. The higher SRB diversity and relative abundance within the lactate-fed reactors may represent an important advantage in terms of process resilience compared to that within the acetate-fed reactor.

Several identified OTUs exhibited a preferential attachment onto the carbon microfibers and were detected at high relative abundances within the CF communities compared to planktonic communities. These included OTUs affiliated to *Rhizobium* (OTU 1), Bacteroidetes (OTU 6), and Alphaproteobacteria (OTU 14) across all three reactors, while Thermotogae (OTU 16) and *Chlorobium* (OTU 17) showed preferential attachment in the 2 L lactate-fed and 2 L acetate-fed reactors, respectively. The dominant planktonic OTU affiliated to Bacterioidetes (OTU 8) differed from that in the attached community (Bacteroidetes (OTU 6)), clearly indicating the preferential attachment of specific OTUs.

Chlorobium, also known as green sulphur-oxidising bacteria (GSB), are photosynthetic microorganisms that perform sulphide oxidation under anaerobic conditions. Henshaw et al. (1998) used *Chlorobium* in a suspended-growth CSTR for sulphide oxidation and achieved approximately 90% sulphide conversion to elemental sulphur. In the current study, the presence of *Chlorobium* within the 2 L acetate-fed reactor explained the green pigmented biomass observed in the bulk reactor volume. The significance of *Chlorobium* and high abundance of *Dethiosulfovibrio* suggests that the conversion of sulphur species within the hybrid LFCR may be more complex than originally thought. Based on the reactor performance it is unlikely that their contribution had a significant impact on the overall performance.

7.3.3.2 Microbial community analysis of the floating sulphur biofilm

The microbial communities in the FSB are, distinctly different to the planktonic (PV and PS) and attached (CF) communities in the bulk reactor volume, as shown by the multivariate PCoA (Figure 7.2) and phyla-class level composition (Figure 7.3). Further analysis of the FSB communities at OTU level is shown in Figure 7.8.

The FSB communities comprised a taxonomically diverse microbial population predominantly represented by known SOB classes Alphaproteobacteria, Betaproteobacteria, Epsilonproteobacteia, Gammaproteobacteria and Chlorobi (Ghosh and Dam, 2009; Tian et al., 2017). In both lactate-fed reactors (Figure 7.8 A and B) at a 5 day HRT, >60% of the mature FSB was comprised of *Rhizobium* (39±12%), *Pannonibacter* (15±11%), *Parracoccus* (12±9%) and *Halothiobacillus* (1.8±0.4%). After the exposure to a 2 day HRT, the shift in microbial community composition of the newly formed FSB in the lactate-fed reactors resulted in the reduction of *Rhizobium* (10.8±0.5%), *Pannonibacter* (8.6±9.3%), and *Parracocus* (6.0±5.0%) with an observed increase in the proportion of *Halothiobacillus* (6.4±0.5%). This was accompanied by an increase in the abundance of OTUs affiliated with the phyla Deltaproteobacteria, Bacteroidetes and Firmicutes dominant within the reactor volume derived communities (CF, PV and PS). The most pronounced increase was exhibited within the 8 L lactate-fed reactor in which the OTUs affiliated with *Desulfovibrio* (12.2%), *Bacteroidetes* (4.4%), *Veilonella* (12.6%) and *Ruminococcacaea* (10.8%) accounted for approximately 40% of the newly formed FSB microbial community.

In the 2 L acetate-fed reactor (Figure 7.8), *Rhizobium* (10.2%) *Pannonibacter* (31.4%) *Acidovorax* (29.2%) *Parracoccus* (0.5%) and *Halothiobacillus* (0.1%) made up 71.4% of the total microbial community in the mature FSB at a 5 day HRT. Similarly, within the 2 L acetate-fed reactor there was a shift in the community structure of the newly formed FSB at a 2 day HRT. This resulted in a decrease in the abundance of *Pannonibacter* (17%) and *Acidovorax* (6.7%) with an increase the proportion of *Rhizobium* (22%), *Parracoccus* (2.7%) and *Halothiobacillus* (5.5%).

Both *Rhizobium* and *Pannonibacter* are classified under Alphaproteobacteria within the order Rhizobiales and Rhodobacteriales, respectively. Tian et al. (2017) showed that the *SoxB* gene is highly conserved within these families. El-Tarabily et al. (2006) reported sulphide oxidation to elemental sulphur by two *Rhizobium* species isolated from calcareous sandy soils. However, *Rhizobium* are generally recognised as nitrogen fixing bacteria. Bai et al. (2019) reported high nitrogen removal rates by a *Pannonibacter* species when fed with ammonia, nitrate or nitrite as a sole nitrogen source. Given the dominance of these two genera within the FSB, it is highly likely that they play an important role in the functioning of the FSB.

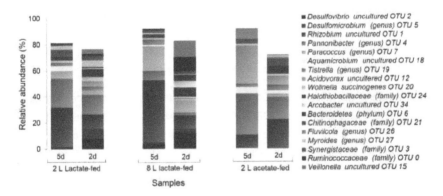

Figure 7.8: Microbial composition at OTU level showing the structural shift in relative abundance within the FSB communities as a function of decreasing HRT from 5 to 2 days in the A) 2 L lactate-fed B) 8 L lactate-fed and C) 2 L acetate-fed reactors

Other SOB genera that were dominant within the FSB included: *Aquamicrobium, Tistrella, Acidovorax, Arcobacter, Wollinella* and *Halothiobacillus*. The most recognised SOB genera, known for their ability to oxidise reduced sulphur, are affiliated with *Halothiobacillus* and *Paracoccus*. Several studies have evaluated the potential application of different SOB genera for treating sulphide-rich waste streams. Vikromvarasiri et al. (2017a) achieved sulphide removal efficiency of 95-100% within a biotrickling filter system inoculated with a strain of *Halothiobacillus neapolitanus*. Vikromvarasiri et al. (2017b) evaluated sulphide removal by a

Parracoccus pantotrophus strain within a biotrickling filter system and achieved removal efficiency of 96% at an initial concentration of 150-400 ppm.

Acidovorax, dominant in the acetate-fed reactor, is generally recognised for catalysing iron oxidation (Carlson et al., 2013). Nalcaci et al. (2011) also reported the potential of an *Acidovorax* species in denitrification. In the 2 L acetate-fed reactor after decreasing HRT from 5 to 2 days, *Acidovorax* decreased sharply from 29 to 7%; it was not detected in either lactate-fed reactors. 16S rRNA sequencing revealed that the FSB was comprised of a diverse SOB community which was distinctly different across the source electron donor. Unexpectedly, the most dominant OTUs were not recognised for sulphide oxidation; however, implicated in nitrogen metabolism. It is well established that the nitrogen and sulphur cycles are closely regulated, with nitrogen utilising bacteria and SOB communities often found inhabiting the same environment (Li et al., 2012). In addition, most SOB can use nitrate as an alternative electron acceptor when oxygen is not available.

As previously discussed, the microbial composition of the FSB observed in the current study was more a reflection of the biofilm maturity than the effect of HRT. Analysis of the FSB communities at OTU level suggest that during initial biofilm formation, a high proportion of SOB are present. As time progresses the biofilm becomes dominated by *Rhizobium* and *Pannonibacter*. However, this would need to be confirmed by evaluating the change in microbial community within the FSB over the stages of development.

The chemosensory motile behaviour of SOB is an important characteristic for understanding the formation of the FSB. Most SOB are microaerophilic "gradient organisms" that often migrate to the narrow zone where oxygen and aqueous sulphide overlap. These microorganisms simultaneously consuming oxygen and sulphide, maintaining steep concentration gradients with elevated transport of both substrates. Simultaneous consumption of oxygen and sulphide in combination with chemosensory motility creates a niche environment in which SOB thrive. The establishment of SOB communities within discrete microenvironments have been described within sulphide-rich sediments. This explanation can also be applied to explain the air-liquid interface within the LFCR. Oxygen from the surrounding environment and high sulphide generated by SRB within the bulk volume creates the counter gradients suitable for SOB activity.

Overall, the shift in microbial community composition confirms that the change in HRT had a corresponding effect on the overall microbial community and the observed reactor performance. The fluctuation of key microbial taxa affiliated with sulphate reduction and sulphide oxidation as well as those implicated in fermentation reveal the importance of regulating the HRT. Studies have reported that the change in environmental variables associated with the change in operational parameters such as pH, reoxidation of sulphide to sulphate, decrease in dissolved sulphide and organic carbon contribute to the changing microbial communities and affected reactor performance (Vasquez et al., 2018). In the current investigation, the microbial community was subject to change in substrate loading and HRT during which different physiological traits and metabolic potential were selected among the microbial population. This caused a shift in the relative abundance of the dominant microbial species within the reactor.

7.4 Conclusion

The outcome of this study revealed that the microbial community composition was largely driven using different carbon substrate rather than by reactor geometry and scale up of the hybrid LFCR. Currently, design criteria for inoculum selection do not exist for sulphate reducing bioreactors. The application of inoculum design could prove an effective engineering tool for successful process operation and has been previously suggested by Pruden et al. (2007). The potential application of acetate as an alternative carbon source to lactate is dependent on the ability to select for dominant complete oxidising SRB.

The study established a link between microbial community dynamic and the observed reactor performance when changing the dilution rate and associated substrate loading rate. At low dilution rates, SRB had a competitive advantage over fermentative microorganisms. Operation at higher dilution rates resulted in a shift in the metabolic degradation pathway of lactate towards fermentation, resulting in the accumulation of propionate and acetate. These results were substantiated by corresponding shifts in the relative abundance of the microbial communities implicated in sulphate reduction associated with lactate oxidation and in lactate fermentation reactions. This demonstrated the importance of the microbial community in determining the success of the treatment process.

In this study, both quantitative and qualitative biomolecular data analysis using multivariate statistics was applied to characterise the microbial communities responsible for process performance. The sequencing of the 16S rRNA amplicon is, however, limiting, particularly when predicting functionality. The need for multivariate statistical analysis of 16S rRNA coupled with functional genes for comparing complex microbial communities, recommended by Rudi et al. (2007), is recommended as an extension of this study.

The lactate-fed reactors exhibited a similar microbial structure at phylum-class level, while further compositional analysis at OTU level informed a greater understanding of the diversity and complexity of the microbial community dynamics across the reactors. Despite slight differences in microbial composition, selection for a high SRB population within both lactate-fed reactors contributed to the relatively high sulphate reduction performance. In the 2 L acetate-fed reactor, the inability to select for a dominant SRB population capable of complete oxidation may have impacted the lower performance.

The FSB represented a unique ecological environment harbouring a taxonomically diverse SOB community that requires further exploration. Due to the limitations of partial sequencing of the 16S rRNA gene, classification was largely limited to genus level with many only identified at phylum and family level. Furthermore, elucidation of the metabolic function of key OTUs within the hybrid LFCR system was not determined, owing to use of 16S rRNA amplicon sequencing. This, however, does provide great opportunity for further research.

The continued development of molecular approaches and the application of whole genome sequencing is expected to prove beneficial toward further advancement in understanding the biologically mediated processes that govern the performance of the hybrid LFCR. This represents an important step for future studies towards integrating microbial community dynamics with bioreactor design and operation.

Chapter 8

Effect of temperature

8.1 Introduction

The hybrid LFCR has been developed as a semi-passive process, requiring minimal energy input and maintenance, hence operation in the absence of temperature control is preferred, making the process susceptible to diurnal and seasonal fluctuations in temperature. Characterisation of the performance of the integrated process as a function of temperature is important to determine process feasibility for demonstration- and large-scale implementation.

In this study, the effect of temperature on fluid flow was evaluated between 10 and 40°C and that on process performance between 10 and 30°C. It was hypothesised that higher operational temperatures increase overall system performance, provided the temperature remains within the operating window of the microbial consortium. Further, decreased operational temperatures, below the optimal operating window, result in reduced overall performance up until a critical point at which further prolonged decrease in temperature results in significant loss in process performance and eventually system failure.

The main objectives and key questions addressed include:

1. Evaluate the effect of temperature on the fluid dynamics within the LFCR and as a function of reactor geometry and scale.
2. Assess the effect of temperature on process performance, namely biological sulphate reduction and partial sulphide oxidation with sulphur recovery, of the hybrid LFCR.
3. Investigate the effect of temperature on process performance based on reactor geometry and scale.
4. Assess the effect of temperature on system performance as a function of electron donor.
5. Evaluate the hybrid LFCR as a dual reactor system for enhance process performance.

The chapter is structured as a study into the effect of temperature on the fluid dynamics, followed by an investigation into the effect of temperature on process performance across the three reactors and the dual operation of the 2 L acetate-fed reactor.

8.2 Effect of temperature on fluid dynamics

8.2.1 Experimental approach

In Sections 4.3 and 5.3, the hydrodynamic mixing within the LFCR was assessed across both the 2 L and 8 L configurations as a function of HRT to characterise the fluid dynamics within the LFCR. This allowed validation of the findings with respect to the conceptual model described by Mooruth (2013). Understanding the mixing pattern within a reactor system is critical to ensure optimal process performance. Hydrodynamics as a function of operating conditions, including HRT and temperature, provide insights into potential inefficiencies affecting reactor performance such as the development of dead zones or short-circuiting. A poor understanding of reactor hydrodynamics may lead to incorrect interpretation of process performance.

To further characterise the hybrid LFCR performance and ensure its optimal operation, an investigation into the effect of temperature on the fluid dynamics was performed. Since the hydrodynamic mixing profile was conserved across both 2 L and 8 L LFCR designs (Section 5.3), this effect was only evaluated in the 8 L LFCR at a constant flow rate, equivalent to a 2 day HRT, and was performed in triplicate using the same sampling (Section 3.1.3) and analytical (Section 3.2) procedure detailed in Chapter 3. The temperature of the bulk liquid in the reactor was regulated by pumping heated or cooled water from a heated or refrigerated circulator bath through the internal exchanger. Temperature conditions assessed were 10, 15, 20, 30, 35 and 40°C.

8.2.2 Results and discussion

The photographic recording of the fluid mixing patterns for the experimental run at 10°C and 40°C are shown in Figure 8.1 and Figure 8.2, respectively. In all previous fluid dynamic studies conducted at ambient temperature (Section 4.3 and 5.3), a consistent mixing pattern was observed, irrespective of the reactor dimensions and HRT range tested. It is therefore expected that similar results, obtained from the current study, would be observed within the 2 L LFCR configuration.

The distinctive mixing pattern observed in Section 5.3 was preserved during operation between 35 and 15°C, with the initial acid feed sinking to the bottom of the reactor before moving in a horizontal direction toward the effluent port. The mixing regime was governed primarily by advective and diffusive transport. However, the fluid mixing pattern observed at 10°C was distinctly different, with a zone of clearing initially forming near the surface of the bulk reactor volume. This was attributed to the effect of temperature differential on the relative density between the inlet feed and the bulk liquid. After 145 minutes of operation the initial zone of clearing remained confined as a narrow zone across the air-liquid interface (Figure 8.1 A and B). By contrast, complete mixing had occurred by this stage, during the previous study, when operated at ambient temperature. It is also likely that during the initial zone of clearing, at the surface, a portion of the feed was lost to short-circuiting over time. This coupled with the decreased diffusion rate contributed to the increase in mixing times exhibited at the low temperature range (10 and 15°C).

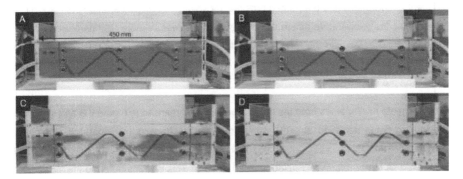

Figure 8.1: Photographic images showing the progression of mixing at 10°C in the hybrid LFCR operated at a flow rate equivalent to a 2 day HRT, photographs taken at A) 34 min B) 120 min C) 208 min D) 235 min.

The experiments conducted at temperatures above ambient showed rapid diffusive mixing during the initial stages of the tracer studies (Figure 8.2). While the majority of the reactor volume was neutralised within a relatively short period of time, a number of 'dead zones' were observed to remain at the base of the reactor. This affected the time required for complete mixing, resulting in longer mixing times to be recorded than previously observed at ambient temperature. This contributed to the large deviation observed across the recorded mixing times at 40°C in the triplicate experimental runs (Figure 8.3).

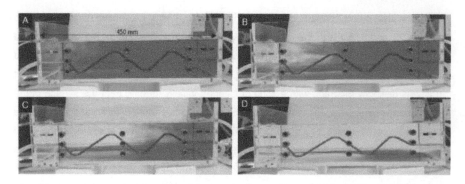

Figure 8.2: Photographic images showing the progression of mixing at 40°C in the hybrid LFCR operated at a flow rate equivalent to a 2 day HRT, photographs taken at A) 88 min B) 115 min C) 145 min D) 172 min

Similar to lower temperatures, there was a temperature differential between the feed (at ambient temperature) and the heated bulk reactor volume. In all cases, attempts were made to minimise the temperature differential by adjusting the feed reservoir temperature to that of the reactor. This reduced the temperature differential, but the slow flow rate resulted in the feed gaining or losing heat on pumping from reservoir to reactor. The alterations had minimal impact on the results.

The complete mixing times from the tracer study are summarised in Figure 8.3 and show that the mixing times achieved at 20°C were similar to the tracer study operated at ambient temperature and a 2 day HRT (2.5±0.05 h) (Section 5.3). As expected, complete mixing times increased as temperature decreased from 20 to 10°C. This was attributed to the slower rate of diffusion experienced at lower temperatures. Unexpectedly, an inverse trend was observed at the higher temperature range, with incremental increases in mixing times recorded at 30, 35 and 40°C. While the increase in complete mixing times were substantial at the low and high end of the temperature range tested, they were still considerably shorter (<4 h) than the operating HRT of 2 days.

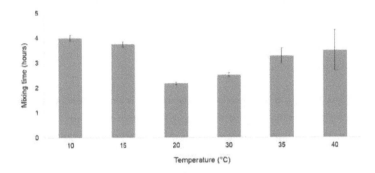

Figure 8.3: Complete mixing times of the dye tracer study as a function of temperature within the 8 L reactor operated at a 2 day HRT. The experimental runs were performed in triplicate with the standard deviation shown by the error bars.

These studies demonstrated the influence of temperature on the fluid dynamics and mixing times in the LFCR. The impact of temperature differential on the density of the inlet feed and bulk volume, at the low and high end temperatures, had major implications on the overall fluid dynamics and mixing times. Therefore, these results must be taken into consideration when further evaluating the effect of temperature on process performance. Since the temperature differential could not be resolved under the experimental set-up, an investigation on the effect of overall density limiting the difference between the feed and bulk reactor volume is recommended as an extension of this study.

8.3 Effect of temperature on process performance

8.3.1 Experimental approach

The effect of temperature on the integrated system was conducted by assessing the performance of biological sulphate reduction and partial sulphide oxidation over 30, 25, 20, 15 and 10°C. Following the HRT study performed in Section 6.3, all reactors were maintained at a 2 day HRT, selected based on optimal system performance at a 2 day HRT. After assessing stable performance at 30°C, the temperature was reduced incrementally by 5°C to a final temperature of 10°C. Temperature was monitored regularly to ensure the desired temperature conditions were maintained within the reactor volume using a digital thermometer. For each temperature condition the system was operated continuously during which biofilm disruption was performed intermittently followed by a biofilm harvest at the end of each experimental run (temperature change). The reactor performance was monitored regularly as described in Section 3.2. The harvested FSB was collected, dried and stored for elemental analysis.

8.3.2 Results and discussion

8.3.2.1 Effect of temperature on process performance

At the start of this study, the 2 L lactate-fed reactor had been operated continuously for 1137 days (>3 years) since the initial demonstration of the proof of concept and HRT studies. The 8 L lactate-fed and 2 L acetate-fed reactors were operated for approximately 428 and 524 days, respectively. Over this period, the carbon microfibers were not removed, thus the extent of biomass colonisation and accumulation within the reactors were significant.

The residual sulphate concentration profiles obtained from the respective reactors are shown in Figure 8.4. The results revealed an increasing trend in sulphate concentration over time as the temperature was decreased. The response to decreasing temperature was consistent across all three reactors. In the 2 L lactate and 8 L lactate-fed reactors, residual sulphate increased from an average of 4.64 and 4.66 mmol/L at 30°C to 8.15 and 8.66 mmol/L at 10°C, respectively. In the 2 L acetate-fed reactor, the concentration increased from 6.44 to 9.52 mmol/L. These results were consistent with the hypothesis that sulphate reduction activity decreases with temperature, resulting in reduced conversion. Overall, based on the residual sulphate concentration data, both 2 L and 8 L lactate-fed reactors exhibited similar performance across the range of temperatures evaluated, while the 2 L acetate-fed reactor was less efficient in sulphate conversion.

As observed in Section 6.3, biofilm disruption impacted performance with sulphate concentration increasing on FSB collapse, then decreasing to a pre-perturbation concentration as FSB re-established. The recovery period following biofilm disruption within the acetate-fed reactor was much longer than the lactate-fed reactors. The sudden decline in acetate observed on decreasing the temperature to 25°C may be less a function of the change in temperature than a delayed performance recovery period. As a consequence, the 2 L acetate-fed reactor was operated at 25°C for an extended period to re-establish stable performance, before decreasing to 20°C.

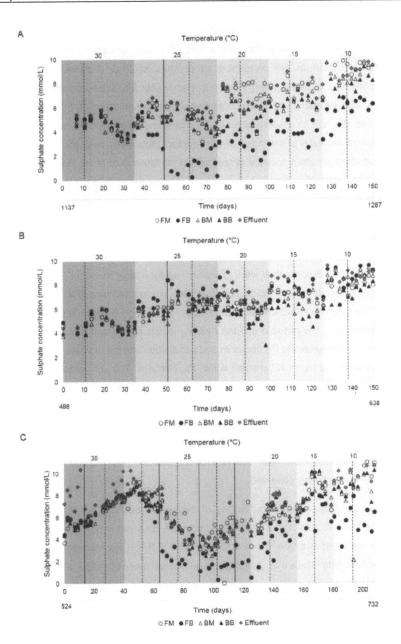

Figure 8.4: Effect of temperature on residual sulphate concentration measured in the reactor (FM, FB, BM and BB) and effluent samples over time in the A) 2 L lactate-fed, B) 8 L lactate-fed and 2 L acetate-fed reactors. Biofilm disruption and harvest events are indicated by vertical dotted and solid lines, respectively. A change in temperature is indicated by the transition in shading intensity and was accompanied by a biofilm harvest.

Prior to decreasing the operational temperature, minimal variation across reactor sampling ports (FM, FB, BM, and BB) were observed at 30°C. This was consistently maintained across all reactors since start-up. On decreasing temperature to 25°C, a distinct divergence in sulphate distribution across sampling ports occurred in both 2 L lactate-fed and 2 L acetate-fed reactors. The greatest variation was observed within the sampling port FB. The decrease in temperature also had an adverse effect on the planktonic cells within the reactors, with the accumulation of biomass at the base of the reactor, particularly near the inlet port. The biomass appeared to form a colloidal suspension in the lower third of the reactors, shown in Figure 8.5. The decrease in temperature may have affected the mobility of the planktonic cells within the bulk volume of the reactor. This coupled with the limited turbulence in the LFCR likely facilitated the accumulation of biomass at the bottom of the reactor. The settled biomass accumulated over a period of time to just beyond the height of the sampling port (FB), in both 2 L lactate-fed (Figure 8.5 A) and 2 L acetate-fed (Figure 8.5 B) reactors. When samples were withdrawn from the FB ports, they were characterised by a large amount of biomass and a lower residual sulphate concentration than the other reactor samples (FM, BM, BB). This suggested that sulphate reduction was more efficient within the FB zone, due to the high cell density. Alternatively, it is possible that the hydrodynamics shifted and a portion of the feed bypassed the zone adjacent to FB due to limited penetration into the biomass. This may have resulted in a localised high HRT present within the biomass. The accumulation of biomass was less pronounced at the far end of the reactor, near the effluent port, with less deviation observed between samples BM and BB, irrespective of the operating temperature. The distinct trend observed in the FB sampling port was not present within the 8 L lactate-fed LFCR in which the accumulation of biomass fell just below the height of the FB port. Consequently, the sample (FB) was consistent with the rest of the reactor and maintained the uniform concentration as previously observed.

Interestingly, the disruption and harvesting of the biofilm over the duration of the study impacted the residual sulphate concentration less than observed in previous investigations. Overall the dissolved sulphide concentration (Figure 8.6) profiles corresponded with the residual sulphate data (Figure 8.4). As temperature was decreased from 30 to 10°C, there was a gradual decrease in the average sulphide concentration.

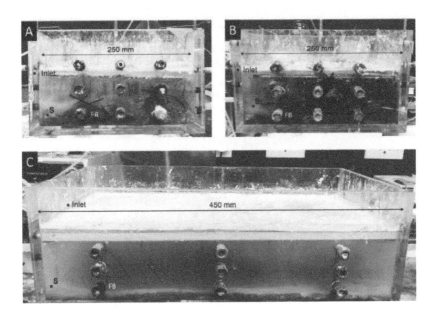

Figure 8.5: Photographs taken during the temperature study showing the A) 2 L lactate-fed (day 126), B) 2 L acetate-fed (day 91) and C) 8 L lactate-fed (day 126) reactors. The symbol (S) highlights the region of sludge build-up near the inlet port and position of the sampling port FB. Internal length dimensions are shown for scale.

The vertical stratification of biomass accumulation at the bottom of the reactor is more clearly observed within the sulphide data (Figure 8.6 A). This was highlighted by the difference in concentration observed in the bottom (FB and BB) and top (FM and BM) sampling ports. This stratification was most pronounced within the 2 L lactate-fed reactor. Since the hydrodynamic study showed that the mixing times were considerably lower than the operating HRT (2-4 h vs 2 days), even at low temperatures and the observed stratification only occurred during the current study, the sulphide stratification between reactor sampling points is likely a function of spatial differences in the rate of sulphate reduction, or the formation of "dead zones", where the local HRT within the biomass is longer than the mean HRT across the reactor. This may be a consequence of the substantial accumulation of planktonic biomass over the extended operation and settling in response to density differences, reduced metabolic activity and motility at reduced temperature or low mixing or a combination. This would explain the less pronounced stratification observed in the 8 L lactate-fed and 2 L acetate-fed reactors which had been operated for a shorter time period, compared to that of the 2 L lactate-fed reactor.

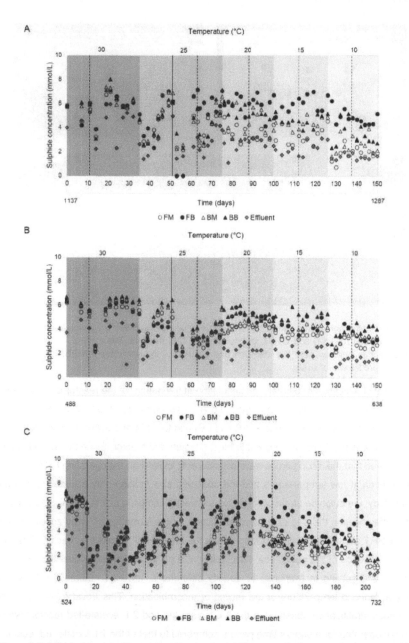

Figure 8.6: Effect of temperature on the dissolved sulphide concentration over time measured in the reactor (FM, FB, BM, and BB) and effluent samples over time in the A) 2 L lactate-fed, B) 8 L lactate-fed and C) 2 L acetate-fed reactors. Biofilm disruption and harvest events are indicated by vertical dotted and solid lines, respectively. A change in temperature is indicated by the transition in shading intensity and was accompanied by a biofilm harvest.

On average the effluent sulphide concentration was consistently lower than the reactor samples (FM, FB, BM, and BB) throughout the duration of the study (Figure 8.6). This was a strong indication that a portion of the sulphide had been oxidised at the interfacial zone. This, together with the minimal variation in sulphate concentration observed between the reactor samples and effluent (Figure 8.4), indicated that the sulphide was partially oxidised to elemental sulphur.

The effect of temperature on pH is presented in Figure 8.7 as the average reactor samples and effluent pH. The pH remained relatively stable over the duration of the study. However, there were distinct differences observed across reactor systems. In the 2 L lactate-fed and 8 L lactate-fed reactors, the average pH ranged between 7.5 and 7 within the reactor samples with an elevated pH observed within the effluent samples. For the majority of the study the 8 L lactate-fed reactor maintained a pH equivalent to the feed (pH 7) when operated between 20 and 10°C, with minimal fluctuation even after biofilm disruption and harvesting. This was distinctly different to previous experiments where the reactor pH sharply increased after biofilm disruption and coincided with a rapid decrease in sulphide concentration. The pH profile measured in the 2 L acetate-fed reactor (

Figure 8.7 C) was higher than in both lactate-fed reactors. This can be attributed to the oxidation of acetate as a carbon source generating more alkalinity than the partial oxidation of lactate. In addition, the partial oxidation of lactate generates a mole of acetate for every mole of lactate consumed with the residual acetate concentration typically above 10 mmol/L, thus lowering the pH within the lactate-fed reactors. The effluent pH was consistently higher than the reactor samples within all three reactor systems. This was attributed to partial sulphide oxidation at the surface, where hydroxyl ions are released during the formation of elemental sulphur. These findings corresponded well with the sulphate and sulphide data. In addition, it also confirmed that the complete oxidation to sulphate was unlikely, as this would have resulted in a decrease in pH due to the formation of sulphate (sulphuric acid).

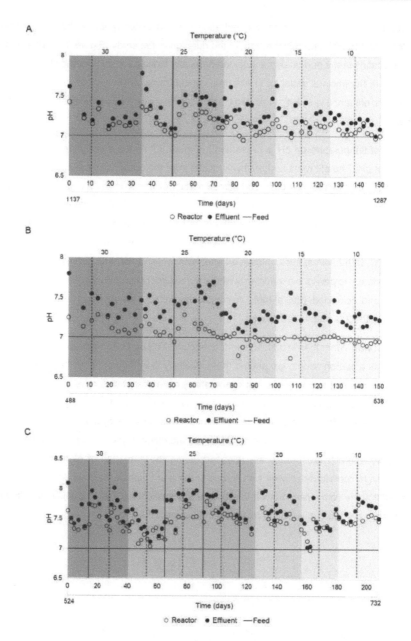

Figure 8.7: Effect of temperature on average pH measured in the reactor (FM, FB, BM, and BB) and effluent samples over time in the A) 2 L lactate-fed, B) 8 L lactate-fed and C) 2 L acetate-fed reactors. Biofilm disruption and harvest events are indicated by vertical dotted and solid lines, respectively. A change in temperature is indicated by the transition in shading intensity and was accompanied by a biofilm harvest.

As previously described, the total sulphide conversion was estimated based on the cumulative difference between the expected sulphide concentration and final effluent concentration over the duration of each experimental run. At 30°C, the sulphide concentration, after biofilm disruption, rapidly decreased to a minimum before increasing. Though this is a well-established characteristic observed in the hybrid LFCR, the increasing and decreasing trend in sulphide concentration was less pronounced compared to the results observed in Section 5.4. This becomes more apparent in the current study, once operated at 20°C where the sulphide concentration became relatively stable with a consistent portion of sulphide conversion between the reactor and effluent samples, as shown in Figure 8.8. These results may indicate possible reduction in sulphide oxidation due to the colder region of the bulk volume with sulphide oxidation favoured at the surface where the effluent port is situated. The sulphide oxidation kinetics as a function of temperature is explored in greater detail within Section 8.3.2.5 Sulphide oxidation kinetics In 2 L acetate-fed reactor, the reoccurring trend in sulphide concentration was exhibited during the operation at 30 and 25°C (Figure 8.8 C). However, the fluctuation in sulphide concentration was most pronounced during the extended period at 25°C between biofilm disruption events. In contrast to the lactate-fed reactors, the difference between the reactor samples and effluent were less significant.

The VFA concentration profiles for the respective reactors were determined to evaluate the effects of temperature on carbon source metabolism and utilisation (Figure 8.9). During operation at 30°C, in both 2 L and 8 L lactate-fed reactors (Figure 8.9 A and B), complete lactate utilisation was observed. The high accumulation of acetate concomitant with the conversion of sulphate indicated that partial oxidation of lactate toward sulphate reduction was the dominant metabolic pathway. This also suggested that the complete oxidation of lactate was not favoured under the operating conditions. In the 2 L and 8 L lactate-fed reactors, while operated at 30°C, the residual acetate concentrations remained relatively stable measuring at 11.2 and 11.5 mmol/L, respectively. In addition, a consistent production of propionate (± 1.8 mmol/L) was observed in both lactate-fed reactors. This indicated that a portion of the lactate was metabolised via fermentation.

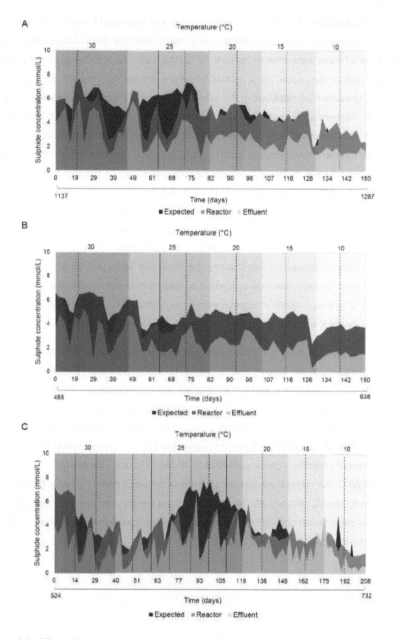

Figure 8.8: Effect of temperature on the expected sulphide generated and sulphide measured in the reactor (FM, FB, BM and BB) and effluent samples as a function of temperature over time in the A) 2 L lactate-fed, B) 8 L lactate-fed and C) 2 L acetate-fed reactors. Biofilm disruption and harvest events are indicated by vertical dotted and solid lines, respectively. A change in temperature is indicated by the transition in shading intensity and was accompanied by a biofilm harvest.

The exposure to decreasing temperature from 30 to 10°C resulted in a shift in the VFA concentration profile over time (Figure 8.9). At 25°C there was no change in VFA concentration up until day 60 (Figure 8.9 A and B) where an increase in residual lactate accompanied by a decrease in acetate concentration was observed. Interestingly, during this period there was minimal change in propionate concentration which suggests that the SRB were more susceptible to low temperature than the fermentative microorganisms at 25°C. This was supported by the decrease in sulphate conversion within both lactate-fed reactors. A further decrease in temperature from 20 through 15 to 10°C resulted in both reduced acetate and propionate concomitant with an increase in lactate concentration over time. By the end of the study at 10°C, the lactate concentration had increased to 3.5 and 5.5 mmol/L while acetate concentrations decreased to 7.7 and 8.2 mmol/L within the 2 L and 8 L lactate-fed reactors, respectively. In addition, the propionate concentrations fell below the detection limit, indicating no fermentation activity. During the operation at low temperature, both SRB and fermentative microorganisms were affected.

The competition between fermentative microorganisms and SRB with lactate as the primary carbon source in sulphidogenic bioreactors is well described within literature (Oyekola et al., 2012; Bertolino et al., 2012). The interaction within the hybrid LFCR was evaluated and discussed as a function of HRT in Sections 4.4.2, 5.4.2 and 6.3.2. Under limiting substrate conditions, SRB outcompete fermentative microorganisms due to their high affinity for the substrate (higher K_s value) (Oyekola et al., 2012). In addition, the high sulphide concentration within a sulphidogenic environment inhibits fermenters. Together, these factors play an important role to ensure the selection of a dominant SRB community for optimal process performance.

While operated at 10°C, the lactate utilisation, on average, equated to 73±14% and 71±16% in the 2 L and 8 L lactate-fed reactors, respectively. The decrease in lactate utilisation was indicative of incomplete substrate utilisation during which both SRB and fermenters were unable to completely metabolise the available lactate. This suggests that the operation at lower temperatures (<15°C) impacted the metabolic rate.

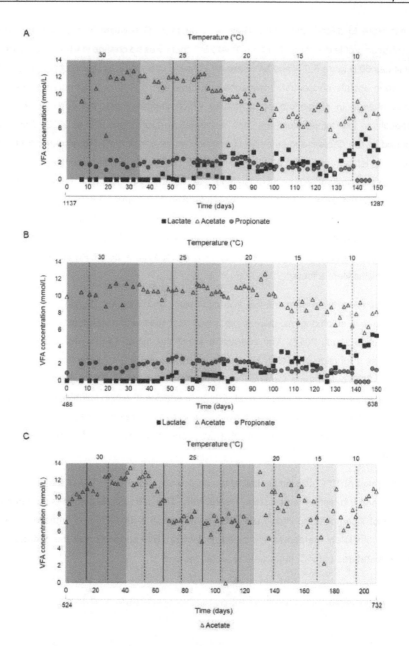

Figure 8.9: The effect of temperature on the VFA concentration profile as a function of time in the A) 2 L lactate-fed, B) 8 L lactate-fed and C) 2 L acetate-fed reactors. Data represents mean values based on reactor sampling ports (FM, FB, BM, and BB). Biofilm disruption and harvest events are indicated by vertical dotted and solid lines, respectively. A change in temperature is indicated by the transition in shading intensity and was accompanied by a biofilm harvest.

During operation at 30°C, in the 2 L acetate-fed reactor, the disruption and harvesting of the biofilm on day 14 resulted in an increase in acetate concentration (Figure 8.9 C) which also corresponded with an increase in residual sulphate concentration (Figure 8.4 C). By day 37, the acetate and sulphate concentrations reached a high of 12.3 and 8 mmol/L, respectively. This was consistent with previous results where the acetate-fed reactor was more sensitive to operational perturbation and required a longer recovery period to regain sulphate reduction performance. As a result, after decreasing the temperature to 25°C the 2 L acetate-fed reactor was operated for an extended period, to allow the system to recovery performance. Once operated at 25°C, there was a decrease in acetate concentration which was accompanied by a decrease in residual sulphate concentration. Over the period of approximately 47 days of operation the acetate and sulphate concentration decreased to 7.3 and 2.5 mmol/L, respectively. A further decrease in temperature from 25 to 10°C in the 2 L acetate-fed reactor resulted in an increase in average acetate concentration.

The volumetric reaction rates are summarised in Figure 8.10. The profiles show the temperature dependency of substrate utilisation and sulphate reduction rate. The response to temperature was distinctly different from that observed when evaluating the effects of HRT (Figure 6.7). A decreasing trend in the volumetric substrate utilisation and production rates was observed as the temperature was reduced from 30 to 10°C in both the 2 L and 8 L reactors which shared a similar profile across volumetric rates. The decrease in both acetate and propionate, concomitant with sulphate reduction rate, indicated that the activity of both SRB and fermentative microorganisms was affected. In the 2 L acetate-fed reactor, there was a sharp decrease in VSRR as temperature decreased, however, the acetate utilisation rate was highly variable due to the high acetate concentrations measured during operation at 30 and 20°C.

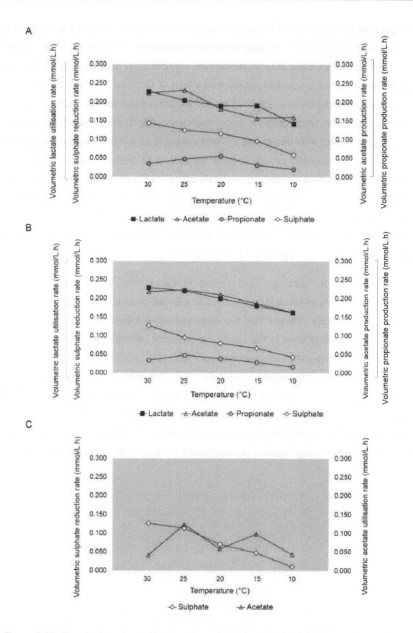

Figure 8.10: Temperature dependency on the volumetric sulphate reduction, substrate (lactate and acetate) utilisation and production (acetate and propionate) rates in the A) 2 L lactate-fed, B) 8 L lactate-fed and C) 2 L acetate-fed reactors.

8.3.2.2 Stoichiometric dependency on temperature

The theoretical and experimental stoichiometric ratio over the duration of the temperature study of the 2 L and 8 L reactors is summarised in Table 8.1 and 8.2. The results revealed an increase in L:S with decreasing temperature. In the 8 L reactor, this ratio increased from 1.8 (where 0.67 represents complete oxidation) gradually to 3.8 (where 2.0 represents incomplete oxidation) as the temperature was decreased from 30 to 10°C. The ratio exceeding 2.0 suggests the utilisation of lactate in competing reactions; however, this is not supported by evidence of fermentation through increased propionate production. The shift in L:S ratio was less pronounced within the 2 L reactor, with experimental ratios remaining constant at 1.6 from 30 to 20°C, then increasing to 2.4 at 10°C, supporting dominant incomplete oxidation under these conditions. The difference may reflect the differing biomass concentrations accumulated in the reactors over their operation of 1137 days for the 2 L reactor compared with 488 days for the 8 L reactor, prior to the initiation of this study. The L:A ratio was consistently stable over the range of temperature (30 to 10°C) and agreed well with the theoretical value for incomplete lactate oxidation (Reaction 8.2). The carbon balance (Effluent: Influent) ranged between 0.8 and 1.0 within both 2 L and 8 L reactors which indicated that most of the carbon added as lactate was accounted for with little other present.

Table 8.1: Theoretical stoichiometric ratios of the metabolic reactions associated with the respective electron donors.

Reaction No.	Chemical reaction	Theoretical stoichiometric ratio		
		L:A	L:S	A:S
8.1	$2\ Lactate^- + 3\ SO_4^{2-} \rightarrow 6\ HCO_3^- + 3\ HS^- + H^+$	-	0.67	-
8.2	$2\ Lactate^- + SO_4^{2+} \rightarrow HS^- + 2\ Acetate^- + 2\ HCO_3^- + H^+$	1.0	2.0	2.0
8.3	$3\ Lactate \rightarrow acetate + 2\ propionate + HCO_3^- + H^+$	3.0	-	-
8.4	$Acetate^- + SO_4^{2-} \rightarrow HS^- + 2\ HCO_3^-$	-	-	1 [a]

[a] acetate utilised per mol sulphate reduced

Table 8.2: Temperature dependency of molar ratio of lactate utilised to moles of acetate and propionate produced involved in biological sulphate reduction, using lactate as the sole carbon-source and electron donor. Average values of experimental stoichiometric ratio are compared with the theoretical ratios (Table 8.1). The carbon balance of total moles VFA accounted compared with the total amount of lactate fed is also presented.

Temperature (°C)	Volumetric rates (mmol/L.h)				Stoichiometric ratios			Carbon balance
	Lactate utilisation rate	Acetate production rate	Propionate production rate	Sulphate reduction rate	Total moles lactate used/mole acetate produced (L:A)	Total moles lactate used/mole sulphate reduced (L:S)	Total moles acetate produced/mole sulphate reduced (A:S)	Total C moles out/ total C mole lactate fed [a] (Effluent:Influent)
				2 L lactate-fed				
30	0.229	0.224	0.036	0.144	1.0	1.6	1.6	0.8
25	0.205	0.232	0.048	0.125	0.9	1.6	1.9	1.0
20	0.190	0.181	0.055	0.116	1.0	1.6	1.6	0.9
15	0.191	0.157	0.031	0.095	1.2	2.0	1.6	0.8
10	0.142	0.159	0.020	0.059	0.9	2.4	2.7	0.9
				8 L lactate-fed				
30	0.229	0.218	0.035	0.128	1.1	1.8	1.7	0.8
25	0.220	0.222	0.048	0.095	1.0	2.3	2.3	0.9
20	0.200	0.209	0.039	0.080	1.0	2.5	2.6	0.9
15	0.180	0.185	0.029	0.067	1.0	2.7	2.8	0.9
10	0.162	0.162	0.016	0.042	1.0	3.8	3.8	0.9

[a] Carbon balance of the total mol VFA measured to the total amount of mol lactate fed

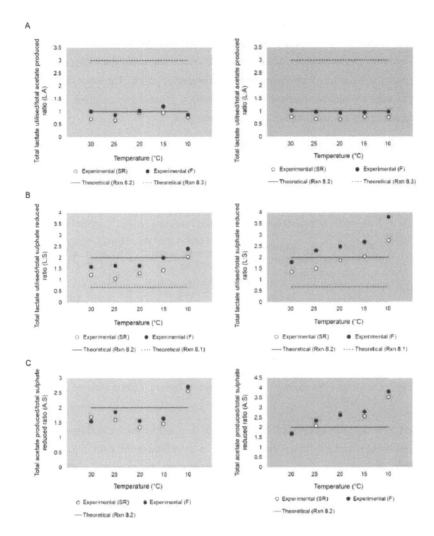

Figure 8.11: Effect of temperature on biological sulphate reduction stoichiometry in the 2 L (left) and 8 L (right) reactors, at a 2 day HRT (dilution rate: 0.0208 1/h), showing the A) Total moles of lactate utilised per mole total acetate produced (L:A), B) total moles of lactate utilised per total moles sulphate reduced (L:S) and C) moles of acetate produced per total moles of sulphate reduced (A:S). Experimental ratio with (F) and without (SR) the contribution of fermentation accounted for, calculated stoichiometrically (Rxn 8.3) based on residual propionate concentration. The horizontal solid (Rxn 9.2) and dotted (Rxn 9.1 and 9.3) lines represent the theoretical ratio for the respective reactions.

The stoichiometry and carbon balance within the 2 L acetate-fed reactor is presented in Table 8.3 and Figure 8.12. The A:S ratio estimated in the 2 L acetate-fed reactor refers the moles of acetate used to sulphate reduced and is theoretically 1.0 for Reaction 4 (Table 8.1). This is distinctly different to that described within the lactate-fed reactor which refers to the total moles of acetate produced to moles sulphate reduced. Over the temperature range 20 to 25°C, the A:S' ratio was near the theoretical value and varied at 30, 15 and 10°C. The most pronounced deviation from the theoretical was observed when operated at a 10°C. This coincided with the decrease is sulphate reduction where only 5% sulphate conversion was achieved (Figure 8.13), suggesting the use of acetate in competing reactions.

Table 8.3: Temperature dependency on volumetric rates and molar ratio of acetate utilised to moles sulphate reduced via sulphate reduction, using acetate as the sole carbon-source and electron donor. Average values of experimental stoichiometric ratios are compared with the theoretical ratios (Table 8.1; Reaction 8.4). The carbon balance of total moles acetate measured to total moles of acetate fed into the system is also shown

Temperature (°C)	Volumetric rates (mmol/L.h)		Stoichiometry	Carbon balance
	Acetate utilisation rate	Sulphate reduction rate	Total Moles acetate used/ mole sulphate reduced (A:S')	Moles of acetate out/ moles of acetate in [a] (Effluent: Influent)
2 L acetate-fed LFCR				
30	0.041	0.127	0.3	1.0
25	0.123	0.113	1.1	0.8
20	0.058	0.070	0.8	0.6
15	0.098	0.047	2.1	0.8
10	0.042	0.010	4.2	0.6

[a] Carbon balance of the total mol acetate measured to the total amount of mol acetate fed

The carbon balance, given as Effluent: Influent lactate ratio, across the 2 L acetate-fed reactor ranged between 0.6 and 1.0. Since YE was found to contribute to the carbon balance with acetate as its breakdown product, it was difficult to evaluate the degree of excess carbon contribution within the system other than the feed acetate concentration. However, considering previous results the lower A:S values were strongly associated with the presence of an additional acetate source, and this is most likely associated with YE and biomass breakdown within the reactor.

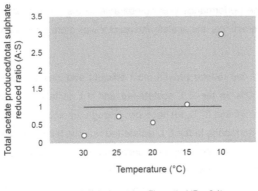

Figure 8.12: Effect of temperature on biological sulphate reduction stoichiometry via Rxn 8.4 in the 2 L acetate-fed reactor, operated at a 2 day HRT (dilution rate: 0.0208 1/h), showing the total moles of acetate utilised per mole sulphate reduced (A:S). The solid horizontal line represents the theoretical ratio based on Rxn 8.4.

8.3.2.3 Biological sulphate reduction kinetics

Results from the study, shown in Figure 8.13, reveal that the decrease in operational temperature, across all three reactors, resulted in a decrease in volumetric sulphate reduction rate (VSRR) (2 L lactate-fed: 0.144 to 0.059 mmol/L.h; 8 L lactate-fed: 0.128 to 0.042 mmol/L.h; 2 L acetate-fed: 0.127 to 0.010 mmol/L.h) and sulphate conversion efficiency (2 L lactate-fed: 66 to 27%; 8 L lactate-fed: 61 to 20%; 2 L acetate-fed: 61 to 5%) over the temperature range between 30 to 10°C. As expected, the highest VSRR and sulphate conversion output was achieved at 30°C. The results agree with previous study that reported an increase in sulphate reduction as temperature increased (Moosa et al., 2002; Al-zuhair et al., 2008). Studies by Greben et al. (2002) and Ferrentino et al. (2017) reported that biological sulphate reduction was stable under temperature perturbations between 20 – 15°C which was found to only account for 3 and 13% decrease in VSRR, respectively. In the current study a 5°C reduction in operational temperature from 25 to 20°C resulted in a 7 and 15% decrease in VSRR in the 2 L and 8 L lactate-fed reactor, respectively. Based on these findings, the 2 L lactate-fed reactor slightly outperformed the 8 L lactate-fed reactor, achieving higher VSRR and sulphate conversion throughout the study. Additionally, the 2 L lactate-fed reactor was less sensitive to temperature perturbation compared to the 8 L lactate-fed reactor, which may be a result of higher relative biomass retention in the 2 L lactate-fed reactor, a consequence of longer operation.

In the 2 L acetate-fed reactor, a sulphate conversion of 61 and 54% was achieved at 30 and 25°C, respectively. The performance was comparable to the conversion observed in the lactate-fed reactors. Interestingly, the acetate-fed reactor was more effective at 25°C than the 8 L lactate-fed reactor.

At 20°C, the 2 L lactate-fed reactor proved most efficient and was capable of maintaining sulphate conversion >50%. In the 2 L acetate-fed and 8 L lactate-fed reactors, sulphate conversion decreased to 34 and 38%, respectively. Though the acetate-fed reactor was able to maintain a similar performance to the 8 L lactate-fed reactor, the sensitivity of the acetate utilising microbial community became apparent once operated at the lower temperature range (15-10°C). A further decrease in temperature affected the sulphate conversion, only achieving 5% within the acetate-fed reactor at 10°C. This was lower than the 27 and 20% conversion achieved in the 2 L and 8 L lactate-fed reactors at 10°C, respectively.

It is encouraging to note that despite the decrease in process performance, sulphate reduction was not completely suppressed when operated at 10°C. From a process perspective the decrease in sulphate reduction rate can be overcome by increasing the HRT. Over the duration of the study, the reactors were operated continuously at a 2 day HRT. Most traditional passive treatment systems are characterised by a longer HRT, often >4 days. A study by Drury (2000) reported that the hydraulic residence time required to achieve 50% sulphate reduction varied from 8 days at 17°C to 41 days at 1°C in an anaerobic solid substrate bioreactor. Alternatively, the addition of a second reactor unit operated in series as a two-stage system can enhance treatment performance.

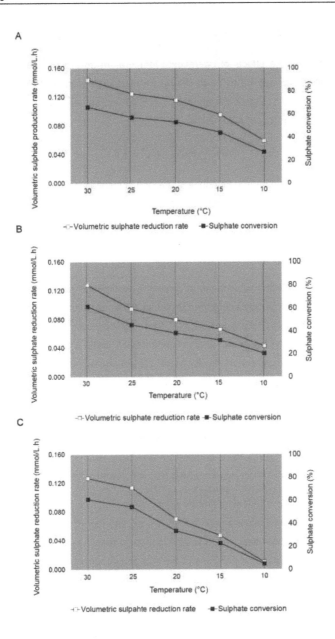

Figure 8.13: Effect of temperature on volumetric sulphate reduction rates and sulphate conversion in the A) 2 L lactate-fed B) 8 L lactate-fed and C) 2 L acetate-fed reactors.

8.3.2.4 Temperature dependence of sulphate reduction rates

To better understand the effects of operational temperature on process performance, the Arrhenius equation was applied to model the temperature dependence of VSRR. Arrhenius curves were generated based on the experimental data and can be represented as follows (Sawicka et al., 2012; Robador et al., 2016):

$$\ln(k) = \ln(A) + \left(\frac{-E_a}{R} \times \frac{1}{T}\right) \qquad \text{(Equation 8.1)}$$

where E_a is the activation energy (J/mol), k is the rate of sulphate reduction (mmol/L.h), A is the Arrhenius constant, R is the gas constant (8.314 J/K.mol) and T is the absolute temperature (K). The activation energies were calculated from the linear range of the experimental data by plotting the VSSR as a function of the inverse temperature (1000/T) (Figure 8.14).

It should be noted that within a biochemical context, E_a estimates acquired from the slope of the linear range are commonly interpreted to reflect the temperature response of the rate-limiting step in a physiological process such as enzymatic catalytic conversions. The activity of an efficient enzyme with low temperature dependence yields a low E_a (Robador et al., 2016). For a mixed microbial community E_a values are not representative of a single sulphate reducing population but reflect the response of the entire complex community. However, studies have shown that calculated E_a values from pure SRB cultures were similar to those of SRB communities derived from marine sediments (Robador et al., 2016). This indicated that many SRB have similar response to change in temperature in pure cultures and complex microbial communities. The E_a value is therefore a suitable estimate that can be applied to describe the temperature sensitivity of the SRB population within the hybrid LFCR process.

In all three reactors, the VSRR-temperature relationship (Figure 8.14) remained linear (2 L lactate-fed: R^2 = 0.98; 8 L lactate-fed: R^2 = 0.98; 2 L acetate-fed: R^2 = 0.96) between 30 and 15°C. As temperature decreased, there is an exponential reduction in the reaction rate, the magnitude of which depends on the value of the activation energy. In the current study, the calculated E_a values were 19.0, 30.6, and 50.4 kJ/mol in the 2 L lactate-fed, 8 L lactate-fed and 2 L acetate-fed reactors, respectively. While the 2 L lactate reactor was the least affected (low E_a) compared to both 8 L lactate-fed and 2 L acetate-fed reactors, the 2 L acetate-fed reactor was the most sensitive (high E_a) to change in temperature.

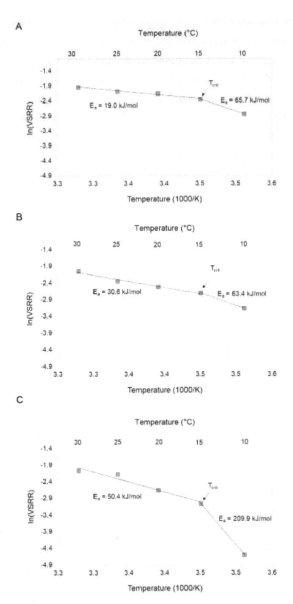

Figure 8.14: Arrhenius plot showing the dependency of volumetric sulphate reduction rate on temperature, showing the E_a values for the linear range between 30 and 15°C as well as the deviation at 10°C in the A) 2 L lactate-fed B) 8 L lactate-fed and C) 2 L acetate-fed reactor. The indicated T_{crit} represents the critical temperature, below which microbial activity is reduced.

Temperatures that deviate outside the linear range of the Arrhenius plot (Figure 8.14) are known as stress inducing temperatures. This was observed when operated at 10°C within all three reactors. As a result the VSRR decreased rapidly, exhibiting a stronger temperature dependency with higher E_a values (2 L lactate-fed: 65.7 kJ/mol; 8 L lactate-fed: 63.4 kJ/mol; 2 L acetate-fed: 209.9 kJ/mol) determined between 15 and 10°C. The temperature at which the two lines (linear range and deviation) intersect is defined as the critical temperature (T_{crit}) (Figure 8.14). The T_{crit} has been proposed for bacterial growth at low temperature to explain the transition between optimal and sub-optimal thermal activity range (Robador et al., 2016). Operation below T_{crit} results in a marked decrease in microbial activity outside of the linear range at which complete inhibition can occur for a given microorganism/s.

The linear range defines the temperature to which a community or microorganism is optimally adapted. For the purpose of the current investigation higher temperatures were not evaluated, thus the T_{max} (maximum temperature at which growth can occur) could not be determined. However, most mesophilic SRB, grow optimally at a temperature between 25 to 40°C (Hao et al., 1996; Moosa et al., 2005; Sheoran et al., 2010). The observed trend in microbial activity-temperature relationship can be explained by direct or indirect effects through which temperature influences the response of an active microbial community. At low temperatures, direct effects can involve decreased growth rate and enzyme activities as well as the alteration of cell composition, while indirect effects are usually associated with the solubility of solute molecules, diffusion of nutrients and osmotic effects on membranes and cell density (Bisht, 2011).

Overall, these results are typical for mesophilic SRB (Sheoran et al., 2010) where the linear range, in the current study, was observed between 30 and 15°C. In addition, these findings highlighted the sensitivity of the SRB microbial community when adapted on acetate as a carbon. This was primarily linked to the slower growth rate of acetate oxidisers (doubling time 10-16 h) compared to lactate oxidisers (doubling time 3-10 h) (Celis et al., 2013). In addition. It has been reported that acetate selects for a less diverse SRB community, since only a few known SRB genera are known to degrade acetate. This would likely infer less robustness within the SRB community present in the acetate-fed LFCR, particularly when exposed to extreme changes in operational conditions.

8.3.2.5 Sulphide oxidation kinetics

Kinetic analysis based on sulphide oxidation performance as a function of temperature is presented in Figure 8.15. The VSOR was calculated based on the theoretical sulphide concentration (expected) derived from the sulphate conversion. The VSOR was determined based on the average sulphide concentration measured within the reactor samples and final effluent. Since the sulphide oxidation within the hybrid LFCR was highly variable due to intermittent biofilm disruption and formation, the VSOR was presented as a function of time, rather than HRT. During the initial stages of the investigation, after a biofilm disruption event was performed, a rapid decrease in dissolved sulphide concentration was observed. As the biofilm regenerated at the surface the dissolved sulphide increased. This distinctive trend was consistent with previous results and was observed within all three reactors during operation at 30 and 25°C. In the reactor samples, the decrease in sulphide concentration after biofilm disruption coincided with an increase in VSOR, reaching a maximum before decreasing over time as the biofilm regenerated and became oxygen limiting.

However, upon further decrease in temperature from 25 to 10°C, the trend in sulphide concentration after biofilm disruption was not observed (Figure 8.6). Instead it remained relatively stable, gradually decreasing as the temperature decreased. This had implications on the VSOR profile and was particularly evident within the lactate-fed reactors where the corresponding VSOR was minimal in the reactor samples, irrespective of biofilm disruption or harvest events (Figure 8.15). This was clearly different from previous results (Section 5.4.2.4) obtained during the evaluation of the effect of HRT where a consistent increase in VSOR after biofilm disruption was observed. The low VSOR recorded within the reactor samples, even though sufficient sulphide was available, suggests that the sulphur oxidising bacteria (SOB) were affected at the lower range temperatures. This was consistent with the poor formation of the FSB at the air-liquid interface. In addition, comparing residual sulphate (Figure 8.4) and dissolved sulphide concentration profiles (Figure 8.6), particularly at 20 to 10°C, the increase in sulphate was more pronounced than the decrease in dissolved sulphide. This suggests that the decrease in temperature had a greater impact on the sulphide oxidation component than on biological sulphate reduction. If sulphate reduction was more affected, the decreasing trend in sulphide concentration would be more pronounced, as it would be oxidised at a faster rate than its generation.

The VSOR based on the measured effluent sulphide was consistently higher than that observed in the average reactor samples and were more variable (Figure 8.15). This was expected since the effluent port is situated at the air-liquid interface where the oxidation of sulphide occurs. At the low range temperatures (20 – 10°C), the VSOR determined in the bulk volume compared to the effluent was considerable, with little oxidation occurring within the bulk volume. The effluent was not affected to the same degree by the decrease in temperature and maintained a relatively high VSOR in both lactate-fed reactors (Figure 8.15 A and B).

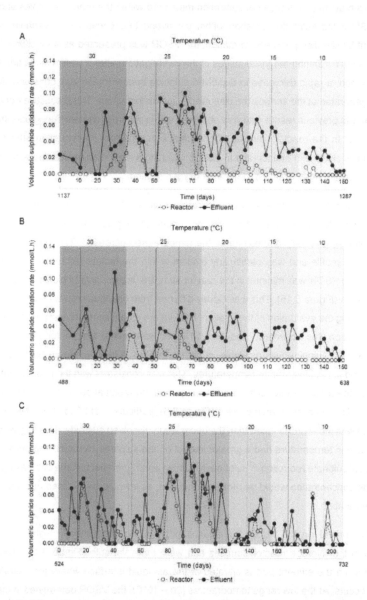

Figure 8.15: Effect of temperature on sulphide oxidation showing the volumetric sulphide oxidation rate and conversion over time in the A) 2 L lactate-fed, B) 8 L lactate-fed and C) 2 L acetate-fed reactors. Biofilm disruption and harvest events are indicated by vertical dotted and solid lines, respectively. A change in temperature is indicated by the transition in shading intensity and was accompanied by a biofilm harvest.

Due to the nature of temperature regulation through the submerged heat exchanger, slight variation in temperature across the reactors was observed. The air-liquid interface, where sulphide oxidation predominantly occurs, was exposed to the controlled temperature in the bulk volume and ambient temperature from the surrounding environment. Consequently, the temperature change at the surface was not as consistent as that within the colder regions of the bulk volume. As a result, a higher VSOR and sulphide conversion was maintained within the final effluent compared to the reactor samples.

In the acetate-fed reactor the distinctive trend observed at 30 and 25°C was more pronounced than compared to the lactate-fed reactors (Figure 8.15 C). In addition, minimal difference in VSOR was observed between the reactor samples and effluent and followed a similar profile in response to biofilm disruption events. Similarly, as the temperature was decreased from 25 to 10°C, in the acetate-fed reactor, there was a decrease in VSOR within the reactor samples, with sulphide oxidation predominantly exhibited within the effluent.

The overall process performance based on sulphate reduction and sulphide oxidation are summarised in Table 8.4. A decreasing trend in average VSOR based on the effluent sulphide was observed in all three reactors as temperature decreased from 25-10°C. This was accompanied by a decrease in sulphate reduction performance. In the hybrid LFCR, the sulphate conversion determines the available sulphide concentration for oxidation to occur. Therefore, it is likely that the sulphide oxidation activity was affected by a combination of decreasing temperature as well as the decline in sulphate reduction performance.

Table 8.4: Summary of the overall reactor performance as a function of temperature. The results highlight the biological sulphate reduction and sulphide oxidation kinetics showing the volumetric rates and corresponding conversion.

Temperature (°C)	Volumetric sulphate reduction rate (mmol/L.h)	Sulphate conversion (%)	Volumetric sulphide oxidation rate (mmol/L.h)		Sulphide conversion (%)[c]
			Maximum[a]	Average[b]	
2 L lactate-fed LFCR					
30	0.144	66	0.077	0.033±0.03	33
25	0.125	58	0.101	0.062±0.03	51
20	0.116	53	0.081	0.045±0.02	44
15	0.095	44	0.069	0.041±0.02	45
10	0.059	27	0.034	0.018±0.01	33
8 L lactate-fed LFCR					
30	0.128	61	0.108	0.038±0.03	31
25	0.095	46	0.064	0.033±0.02	38
20	0.080	38	0.058	0.034±0.01	38
15	0.067	32	0.043	0.029±0.01	38
10	0.042	20	0.042	0.020±0.01	39
2 L acetate-fed LFCR					
30	0.127	61	0.081	0.034±0.03	41
25	0.113	54	0.124	0.044±0.03	43
20	0.070	34	0.059	0.025±0.02	33
15	0.047	23	0.075	0.011±0.02	21
10	0.010	5	0.046	0.009±0.02	23

[a] Maximum VSOR measured in the final effluent
[b] Average VSOR recorded in the final effluent over the duration of the experimental run
[c] Cumulative sulphide conversion based on the expected sulphide and final effluent

8.3.2.6 Biofilm harvest and sulphur recovery performance

Biofilm harvested over the duration of the study as a function of temperature is shown in Figure 8.16. The results showed a decreasing trend in FSB recovery in the 2 L (4.9 – 1.1 g) and 8 L (15.7 – 7.9 g) lactate-fed reactors as temperature decreased from 30 to 10°C. Studies by Sposob et al. (2017) and Xu et al. (2012) investigated the effect of temperature on the removal of sulphide for elemental sulphur production between 25 and 10°C. The studies showed that a decrease in temperature caused a decrease in elemental sulphur recovery. Similarly, in this study, the decrease in temperature resulted in a reduction in the recovery of the sulphur biofilm mass (Figure 8.16). This may suggest that the partial sulphide oxidation through the FSB was similarly affected by decreasing temperature. This was confirmed by the decrease in VSOR and sulphide conversion as temperature decreased. However, in the current investigation, the decrease in sulphate reduction activity was also an influential factor as this determined the availability of sulphide for partial oxidation. Consequently, the decrease in the expected sulphide (sulphate reduced) as temperature decreased limited the maximum sulphur recovery potential through the biofilm. A study by Molwantwa & Rose (2007), evaluated the effects of temperature on the performance of the FSB within a LFCR. The system was operated exclusively for sulphide oxidation and was continuously fed with a constant feed of sulphide. The study reported a decrease in sulphide removal (74-53%) and sulphur recovery (43-28%) when temperature was decreased from 20 to 15°C.

In the 2 L acetate-fed reactor at 25°C, a premature biofilm disruption occurred (Figure 8.17 D). This resulted in the regeneration of the biofilm outside of the studies' parameters of inducing disruption, after a defined period of time. The additionally regeneration cycle of the FSB resulted in a greater mass of biofilm recovered (25 and 20°C) from the 2 L acetate-fed reactor than from the 2 L lactate-fed reactor (Figure 8.16).

At lower temperatures in all three reactors the rate at which biofilm formed at the surface was considerably slower when compared to biofilm formation at 30°C (Figure 8.17). In addition, the structural integrity of the FSB was compromised and is highlighted in Figure 8.17, showing the degree of biofilm formation. After 2 days while operated at 30°C (Figure 8.17 A) a well-established biofilm covered the entire surface of the reactor. In contrast, during operation at 10°C (Figure 8.17 B and C) the biofilm was less developed and had a consistency that resembled the initial sticky/thin stage of formation. The decrease in temperature may have had an adverse effect on the microbial community responsible for the initial development of the biofilm at the surface (EPS producers), coupled with the reduced activity of the SOB population responsible for sulphide oxidation. Consequently, the reactors could not sustain the development of a structurally sound FSB at the lower temperatures. It is no coincidence that the degree of biofilm recovery followed a similar trend to that of the observed decrease in biological sulphate reduction as the temperature decreased.

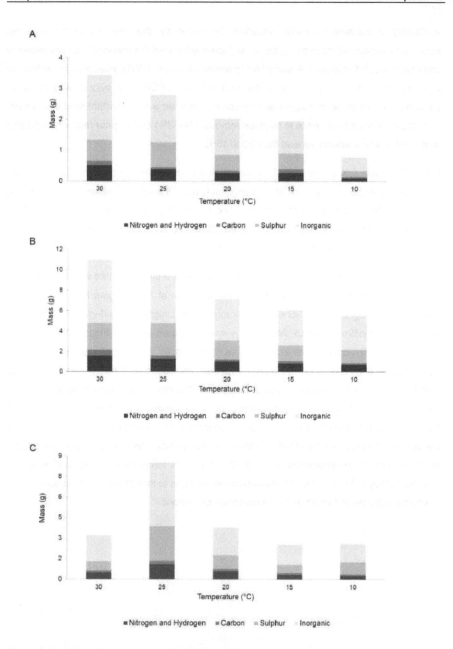

Figure 8.16: Effect of temperature on the floating sulphur biofilm recovery showing the composition of inorganics, sulphur, carbon, nitrogen and hydrogen based on the amount (grams) of biofilm harvested, in the A) 2 L lactate-fed, B) 8 L lactate-fed and C) 2 L acetate-fed reactors.

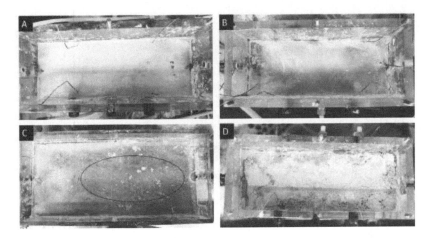

Figure 8.17: Photographs of the floating sulphur biofilm formation after disruption while operated at 2 day HRT showing the A) 2 L lactate-fed reactor at 30°C after 2 days (well-developed biofilm), B) 2 L lactate-fed reactor at 10°C after 2 days (poor formation), C) 8 L lactate-fed reactor at 10°C after 2 days (circle indicates clear zone) and D) 2 L acetate-fed reactor at 25°C after 5 days (premature disruption).

Elemental analysis of the biofilm revealed a shift in composition as temperature decreased (Figure 8.18). Remarkably, there was a 10% increase in sulphur composition when the temperature decreased from 30 to 25°C and was consistent across all three reactors (2 L lactate-fed: 20 to 29%; 8 L lactate-fed: 24 to 34%; 2 L acetate-fed: 21 to 30%). This was accompanied by a decrease in the carbon (C), nitrogen/hydrogen (NH), and inorganic (I) fraction. It is likely that the formation of the inorganic crystals was favoured at 30°C. The 10% increase in elemental sulphur composition was substantial when considering that a further decrease in temperature from 25 – 10°C had less of an impact, only decreasing by 4% and 8% in the 2 L and 8 L lactate-fed reactors, respectively. Previous studies that initially evaluated the potential application of FSB for partial sulphide oxidation revealed that the biofilm predominantly comprised of elemental sulphur (Molwantwa, 2008). Further studies by Mooruth, (2013) reported sulphur content as high as 53 to 94% with the inorganic fraction only accounting for <4.5%. The results obtained from this study was distinctly different where the inorganic fraction accounted for 56±4% in the FSB. In Section 4.3.2.4, it was established based on SEM-EDX analysis that large inorganic crystals, embedded within the FSB, predominantly comprised of magnesium and phosphate. The major contributing factor for the formation of the crystals was attributed to the composition of the modified Postgate B medium, which contained large amounts of magnesium and phosphate. This was not used in previous studies by Molwantwa (2008) and Mooruth (2013) that evaluated the FSB. It is possible that under the operating conditions, evaporative crystallisation was favoured at the surface with the FSB potentially serving as a nucleation site.

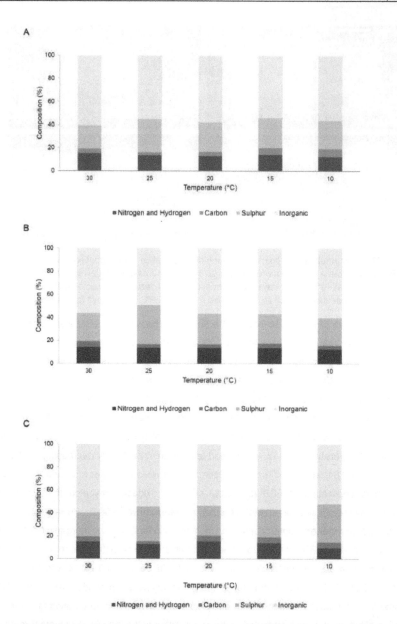

Figure 8.18: Effect of temperature on the elemental composition of the floating sulphur biofilm, showing the composition (%) of inorganics, sulphur, carbon, nitrogen and hydrogen present in the harvested FSB collected from the respective experimental runs (30 to 10°C) in the A) 2 L lactate-fed, B) 8 L lactate-fed and C) 2 L acetate-fed reactors.

The precipitate may have influenced the rate at which the biofilm becomes oxygen limiting, thus affecting overall sulphide removal and sulphur recovery in the FSB. In Section 5.4.2.5, an attempt to reduce the inorganic precipitate by decreasing the amount of magnesium sulphate in the feed proved unsuccessful. Further studies should be applied to characterise the inorganic crystal and to reduce the excess inorganics present in the medium.

A summary of the sulphur recovery through the FSB over the duration of the study is presented in Table 8.5. Overall the FSB-S (grams of sulphur recovered from the biofilm) based on the converted sulphide-S (grams of sulphur) over the duration of the experiment averaged at 30±4.6% and 33±6.0% in the 2 L and 8 L lactate-fed reactors respectively. The unaccounted for fraction of converted sulphide (gap-S) not recovered through the FSB was calculated to be approximately 70%. Considering that there was minimal increase in residual sulphate concentration between the reactor samples and final effluent (Figure 8.4), the complete oxidation to sulphate was negligible. Therefore, the gap-S fraction was predominantly made up of fine colloidal sulphur particles suspended in solution and fragments of biofilm that were released into the effluent over time. This was evident by the build-up of sulphur that was observed within the effluent pipe. In addition, due to the hydrophilic nature of biological produced sulphur, the physical disruption of the FSB between harvesting, may have aided in the dispersion of the sulphur into the liquid phase. The suspension and loss of sulphur to the bulk volume during biofilm harvesting was similarly reported by Molwantwa, (2008) and Mooruth, (2013). The sulphur can be recovered through sedimentation and centrifugation which are standard practices used in many commercial sulphide oxidation processes (Syed et al., 2006; Cai et al., 2017).

In the 2 L acetate-fed reactor the estimation of sulphur recovery based on converted sulphide at 15 and 10°C could not be determined. This was largely due to the low sulphate reduction during which some discrepancies were observed based on the expected and measured sulphide. The regeneration and enhanced recovery of the FSB highlighted the importance of regulating the frequency of biofilm disruption for optimal sulphur recovery.

Table 8.5: Elemental sulphur recovery performance as a function of temperature across all three reactors showing the amount of sulphur recovered from harvesting of the biofilm (FSB-S) as well as the respective recovery (%) based on each parameter including the total sulphate-S load, expected sulphide-S (sulphate reduced) and converted sulphide-S (sulphide oxidised).

Temperature (°C)	FSB-S [a] harvested Amount (g)	Total sulphate-S load Amount (g)	Total sulphate-S load Recovery (%) [b]	Expected sulphide-S Amount (g)	Expected sulphide-S Recovery (%) [b]	Converted sulphide-S Amount (g)	Converted sulphide-S Recovery (%) [b]	Gap-S [c] (g)
				2 L lactate-fed				
30	0.7	12.4	6	6.9	10	2.3	30	1.6
25	0.8	14.2	6	7.4	11	3.8	21	3.0
20	0.5	8.9	6	3.9	13	1.7	30	1.2
15	0.5	9.2	6	3.5	15	1.6	32	1.1
10	0.2	8.5	2	1.8	11	0.6	32	0.4
				8 L lactate-fed				
30	2.7	51.3	5	28.3	10	8.9	30	6.2
25	3.2	58.7	6	25.3	13	9.6	33	6.4
20	1.9	36.7	5	15.4	12	5.9	32	4.0
15	1.5	38.1	4	14.9	10	5.7	27	4.2
10	1.3	35.2	4	7.9	17	3.1	43	1.8
				2 L acetate-fed				
30	0.7	15.2	5	5.6	13	2.3	29	1.6
25	2.5	9.9	17	8.4	20	3.6	47	1.0
20	1.0	9.6	10	2.8	36	0.5	>100	-
15	0.6	6.7	9	2.0	30	-	-	-
10	0.9	15.2	6	0.4	>100	-	-	-

[a] Amount of sulphur recovered from FSB
[b] Sulphur recovered from the FSB based on the respective parameter
[c] Gap-S colloidal and sulphur biofilm released into the effluent

8.3.2.7 Reactor resilience and restoration of process performance

After been exposed to low temperature at 10°C, the 2 L and 8 L lactate-fed reactors and the 2 L acetate-fed reactor had been operated continuously for 1333, 684 and 732 days, respectively. The reactors were then restored to more favourable conditions in order to evaluate the resilience of the reactors to re-establish sulphate reduction performance (Table 8.6). Both lactate-fed reactors were able to recover process performance once operation was restored to a low dilution rate at a 4 day HRT and increased temperature at 25°C. The 2 L and 8 L lactate-fed reactors achieved a sulphate conversion of 61% and 47% within 30 days after restoring operating conditions, respectively. The ability of the 2 L lactate-fed reactor to recover sulphate conversion was attributed to the extensive biomass accumulation which was considerably higher than observed in the 8 L lactate-fed reactor. The ability to recover process performance highlighted the resilience of the microbial population to overcome exposure to stress conditions. The period required for the reactors to recover performance was dependent on the proportion of the active SRB population as well as the degree of biomass retention within the respective reactors. A key parameter for inducing change to a microbial culture to favour a desired characteristic is the period of exposure to a defined set of conditions. It could be argued that an extended period of operation at a 4 day HRT without biofilm disruption, beyond the 30 days, could have facilitated higher sulphate conversion recovery within both reactors over time. Though this was not evaluated, it was demonstrated in Section 5.4.2, where longer operation between biofilm disruptions favoured higher sulphate conversion.

Table 8.6: Sulphate reduction performance recovery in the reactors after exposure to 10°C. The results summarise the VSRR and sulphate conversion achieved before and after the recovery period, once operation was restored to a longer HRT and higher temperature

Temperature (°C)	HRT (days)	Recovery period (days)	VSLR (mmol/L.h)	VSRR (mmol/L.h)	Sulphate conversion (%)
2 L lactate-fed					
10	2	30	0.217	0.059	27
25	4		0.108	0.066	61
8 L lactate-fed					
10	2	30	0.217	0.042	20
25	4		0.108	0.046	43
2 L acetate-fed					
10	2	38	0.217	0.010	5
25	5		0.086	0.062	72

Interestingly, the 2 L acetate-fed reactor, which was most affected by the exposure to 10°C with a sulphate conversion of 5%, was able to recovery high sulphate reduction performance, once the operating conditions was restored to a 5 day HRT and 25°C. The 2 L acetate-fed reactor required just 38 days to recover a sulphate conversion of 72% and a corresponding VSRR of 0.062 mmol/L.h. This was significant, considering that in previous experimental runs (Section 6.3.2) the acetate-fed reactor required approximately 100 days to recover from biofilm disruption. Similarly, as observed within the lactate-fed reactors, the extent of biomass accumulation within the 2 L acetate-fed reactor after 798 days of continuous operation is likely to have played a major role in re-establishing high sulphate reduction. During operation at low temperature, metabolic activity was reduced to a state of dormancy. However, once the operating conditions were more favourable at longer HRT and higher temperature, the microbial activity increased.

8.4 Dual reactor operation in series

8.4.1 Experimental approach

In Sections 4.4, 5.4 and 6.3, during the initial demonstration of proof of concept and HRT studies, it was evident that the performance of the FSB and partial oxidation of sulphide was limited due to the frequency of biofilm disruption. The results were consistent with studies by Mooruth (2013) and Molwantwa (2008), which concluded that for optimal performance, the FSB should be harvested every 2-3 days. However, in the hybrid LFCR process this would have adverse effects on the active SRB community and sulphate reduction performance due to oxygen ingress into the bulk volume. This was demonstrated in Section 5.4, where an extended period between biofilm disruption events facilitated higher sulphate conversion. The FSB is governed by the correct sulphide to oxygen ratio, and while in most sulphide oxidation studies this is controlled by maintaining a high sulphide concentration in the feed, the hybrid LFCR process relies on the generated sulphide through sulphate reduction occurring within the bulk volume of the reactor. Based on these parameters and findings in Section 5.4, the hybrid LFCR has since been operated in favour of maximum sulphate reduction performance, with the frequency of biofilm disruption in the current study only occurring after every 12 days. This was significant considering that the optimal frequency of disrupting the biofilm was determined to be every 2-3 days. Consequently, the overall potential for sulphide removal and sulphur recovery in the hybrid LFCR has been restricted with a large portion of untreated sulphide released within the effluent stream.

Toward the end of Chapter 5 it was concluded that the development of a dual reactor system could provide an effective means of addressing untreated sulphide to enhance sulphur recovery and overall process performance. In addition, a secondary reactor in series would facilitate the treatment of residual sulphate and COD. It is well established that SRB communities are better competitors under substrate limiting concentrations and so their activity would be favoured in scavenging the residual COD within the secondary reactor. It is therefore expected that sulphate reduction can be enhanced by operation of an additional reactor in series whereby the effluent stream of the primary reactor is fed into a secondary reactor unit for effective treatment of the residual sulphate, sulphide and COD. It is envisaged that a higher degree of operational flexibility with respect to effluent quality control will be achieved. The conceptual model of the dual hybrid LFCR process is presented in Figure 8.19.

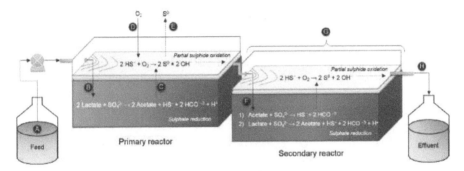

Figure 8.19: Conceptual model of the dual hybrid LFCR process showing the configuration of the reactor setup, where a defined medium (A) is pumped into the primary reactor, (B) biological sulphate reduction occurs within the bulk volume and the generated sulphide (C) is partially oxidised at the air-liquid interface where oxygen (D) is available from the surrounding environment resulting in the formation of a floating sulphur biofilm. The elemental sulphur deposited within the biofilm (E) can be recovered through harvesting. Overflow from the primary reactor is received in the secondary reactor (F) as an additional "polishing stage" where the hybrid process is repeated treating any residual sulphate, sulphide and COD. It is anticipated that the metabolism of the accumulated acetate generated in the primary reactor will serve as the primary electron donor with residual lactate also utilised. Additional (G) FSB recovery from the secondary reactor. The final treated effluent (H) characterised by an enhanced removal of residual sulphate, sulphide and COD concentrations with higher sulphur recovery.

The use of a multi-stage chemostat system, composed of two reactors arranged in series has previously been shown to improve biological sulphate reduction efficiency (Oyekola, 2008). The following section will demonstrate the potential operation of the hybrid LFCR process as a dual system, incorporating two LFCRs in series. In the set-up, the primary reactor was elevated slightly higher in order to facilitate passive flow into the second reactor. Similarly, the secondary reactor (R2) was fitted with carbon-microfibers and a harvesting screen (not shown in Figure 8.20). The dual system operation was set-up on day 69 during the temperature study

Chapter 8   Effect of temperature

described in Section 8.3. Therefore, the results will demonstrate the impact of the secondary reactor on the overall performance of the 2 L acetate-fed LFCR in comparison with the performance presented in the previous section.

8.4.2 Results and discussion

8.4.2.1 Performance of the two-stage LFCR reactor system

The dual LFCR system is shown in Figure 8.20, illustrating the reactor set-up of the second 2 L LFCR (R2) connected to the primary 2 L acetate-fed reactor (R1). The performance data is presented in Figure 8.21. Upon initial start-up, within the first 24 h of operated at 25°C a thin but complete FSB covered the entire reactor surface. A similar trend in residual sulphate concentration (Figure 8.21 A) was observed in both R1 and R2 over the duration of the study, however, there was an increase in sulphate concentration within the secondary reactor (R2). This indicated that a portion of sulphide received from the primary reactor was completely oxidised to sulphate. Furthermore, it suggested that minimal sulphate reduction occurred within the secondary reactor.

Figure 8.20: Dual reactor configuration set-up showing two 2 L LFCRs linked in series. Reactor 1 (R1) was elevated to facilitate passive flow into reactor 2 (R2). A well-developed FSB formed in R2 after just 24 h of initial operation.

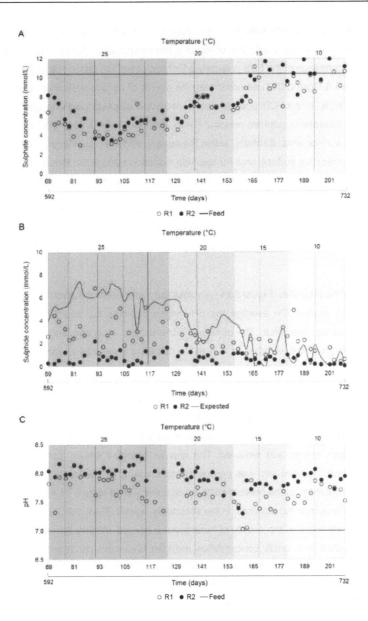

Figure 8.21: Effect of temperature on the performance of the dual reactor system showing the A) residual sulphate concentration, B) dissolved sulphide concentration and C) pH measured over time measured in the effluent samples. Biofilm disruption and harvest events are indicated by vertical dotted and solid lines, respectively. A change in temperature is indicated by the transition in shading intensity and was accompanied by a biofilm harvest.

While theoretically sulphate reduction can occur throughout the bulk volume of the LFCR, sulphide oxidation through the FSB is confined to the surface within the hybrid LFCR process. Thus, the performance of the FSB is largely a function of surface rather than volume. A study by Molwantwa & Rose (2013) demonstrated the effects of surface area on sulphur recovery through the FSB, in a LFCR operated as a dedicated sulphide oxidising unit. The study exhibited an increase in sulphide removal (74 to 88%) and sulphur recovery (43 to 66%) by doubling the surface area. Similarly, within the current study, the additional reactor in series increased the reactive surface area for sulphide oxidation through the formation of the FSB. Thus, resulting in a marked decrease in the dissolved sulphide concentration (Figure 8.21 B).

On average, sulphide concentrations measured in R2 were consistently low, ranging between 0 and 1.2 mmol/L. This resulted in an increase in sulphide conversion during operation at 25°C, from 43% in the 2 L acetate-fed reactor (single reactor) to 81% in the two-stage reactor system.

Monitoring of the pH profile (Figure 8.21 C) across the dual system revealed an increase within the secondary reactor. On average the pH ranged between 8.2 and 7.8 with a gradual decrease observed over time. The observed increase was a consequence of the sulphide oxidation through partial oxidation as well as the decrease in residual acetate concentration.

Analysis of the acetate concentration (Figure 8.22) within the dual reactor system showed a decrease in acetate concentration within the second reactor. The residual acetate measured in the secondary reactor was consistently lower than observed in the effluent of the primary reactor. During initial operation at 25°C, approximately 63% of the incoming residual acetate from the primary reactor was removed. This was an important finding, considering that the reactor was not colonised and was predominantly made up of planktonic cells. In addition, there was no decrease in the residual sulphate concentration (Figure 8.21 A), which indicated that sulphate reduction did not occur in the secondary reactor. Thus, the decrease in residual acetate was attributed to the activity of heterotrophic acetate utilising bacteria. It is possible that the reduction in sulphide concentration provided favourable conditions for non-SRB to thrive since its inhibitory action would be significantly reduced at low concentrations. Furthermore, the exposure to the oxic zone may have inhibited SRB activity during transport from the primary to the secondary reactor.

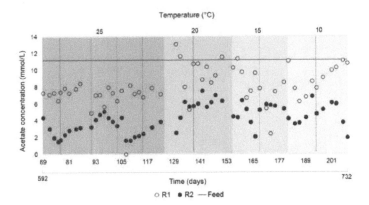

Figure 8.22: The effect of temperature on the acetate concentration profile over time in the dual reactor system in series showing the mean data in R1 and R2. Biofilm disruption and harvest events are indicated by vertical dotted and solid lines, respectively. A change in temperature is indicated by the transition in shading intensity and was accompanied by a biofilm harvest.

8.4.2.2 Biofilm harvest and sulphur recovery

As expected, the addition of a second reactor resulted in an increase in the total amount of harvested FSB (Figure 8.23). Similarly, there was a decreasing trend in the amount of harvested biofilm from R2 as temperature decreased (Figure 8.23 A). The comparison of the elemental composition analysis between R1 and R2 revealed that the Inorganic (I) fraction remained the dominant component in the FSB, accounting for approximately 50 to 60% over the range of temperatures tested. Interestingly, there was an increase in the carbon fraction from 6 to 13%, in the second reactor (Figure 8.23 B), as the temperature decreased from 30-10°C. This was accompanied by a decrease in sulphur content from 30 to 6%, which corresponded with the decline in sulphate reduction performance.

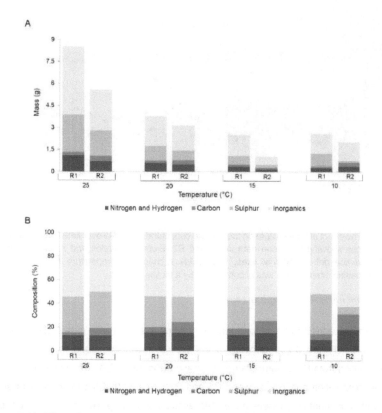

Figure 8.23: Effect of temperature on the elemental composition of the floating sulphur biofilm in the dual acetate-fed reactor system, showing the composition (%) of inorganics, sulphur, carbon, nitrogen and hydrogen present in the harvested FSB collected from the respective experimental runs in primary (R1) and secondary (R2) reactors

An overall summary of the sulphur recovery through the FSB in shown in Table 8.7. A marked increase in sulphur recovery was observed across all parameters evaluated when compared with the results obtained during single-stage operation (Table 8.5). The sulphide conversion (SO) in the dual system ranged between 71 and 82%. The operation at 25°C proved most effective, achieving 82% sulphide conversion and recovering a total of 4.3 g (62%) of elemental sulphur through harvesting the FSB. The converted sulphide, not recovered through the FSB (Gap-S), was present within the effluent as a suspended fraction of colloidal sulphur and fragments of biofilm. This was evident by the accumulation and settling of sulphur within the effluent and reservoir.

The results re-emphasise the importance of maintaining high sulphate conversion in the hybrid LFCR to ensure that the availability of sulphide for partial oxidation to sulphur is not the limiting substrate. As concluded in Section 8.3.2.2, considering the results obtained in Chapter 4, 5 and 6, the overall process performance can be increase through operation at longer HRT to overcome the kinetic constraints experienced at lower temperatures. A longer HRT will favour the activity of SRB and result in higher conversion however this will result in a decrease in VSRR. Therefore, the correct regulation of the operating HRT in the hybrid LFCR is an important parameter, particularly when considering up-scale application where the control of temperature will not be available.

Table 8.7: Overall sulphur recovery performance in the dual reactor system as a function of temperature showing the amount of sulphur recovered from harvesting of the biofilm (FSB-S) as well as the respective recovery (%) based on each parameter including the total sulphate-S load, expected sulphide-S (sulphate reduced) and converted sulphide-S (sulphide oxidised).

Temperature (°C)	FSB-S harvested [a]			Total sulphate-S		Expected sulphide-S			Converted sulphide-S			
	R1 (g)	R2 (g)	Total (g)	Amount (g)	Recovery (%) [b]	SR (%)	Amount (g)	Recovery (%) [b]	SO (%)	Amount (g)	Recovery (%) [b]	Gap-S (g) [c]
						2 L acetate-fed (dual system)						
25	2.6	1.7	4.3	15.2	28	54	8.4	51	82	6.9	62	2.6
20	1.0	0.7	1.7	9.9	17	34	2.8	59	71	2.0	82	0.3
15	0.6	0.2	0.8	9.6	9	23	2.0	41	75	1.5	56	0.7
10	0.9	0.1	1.0	6.7	15	5	0.4	>100	75	0.3	>100	-

[a] Amount of sulphur recovered from FSB
[b] Sulphur recovered from the FSB based on the respective variable
[c] Gap-S colloidal and sulphur biofilm released into the effluent

8.5 Conclusions

This study confirmed that temperature plays a critical role in the overall activity of the sulphate reducing and sulphide oxidising components in the hybrid LFCR process. As temperature was incrementally decreased from 30 to 10°C, sulphate reduction decreased. Based on these findings, at lower temperatures the system may require operation at a longer residence time to compensate for the observed loss in performance. The reduction in performance at low temperature was more pronounced within the acetate-fed system. At 20°C, the decrease in biological sulphate reduction and poor biofilm formation affected the stability and robustness of the 2 L acetate-fed reactor. However, it was observed that the increased frequency of biofilm disruption (prematurely) due to the poor formation of a structurally sound FSB resulted in a higher accumulation and recovery of biofilm. This re-instated the importance of regulating FSB harvesting to facilitate optimal sulphur recovery.

The lactate-fed reactors exhibited a greater resilience when exposed to low temperatures. Comparison of geometry and scale-up as a function of temperature on process performance revealed that the 2 L lactate-fed reactor was more effective. However, this was directly linked to the relative biomass accumulated over the longer period of operation. The data analysis across both reactors were consistent and exhibited a similar response to change in temperature. The results build on Chapters 5 and 6 which demonstrated that aspect ratio and scale up had minimal impact on over process performance.

The dual reactor operation in series proved effective for sulphide removal and sulphur recovery. Though the intention was to treat the residual sulphate and COD, minimal reduction in sulphate concentration was observed in the secondary reactor. This was directly associated with the degree of colonisation. It was anticipated that the overflow received from the primary reactor would contain active SRB that would effectively colonise the secondary reactor. However, given the conditions within the reactor, the SRB were likely outcompeted by heterotrophic acetate utilising bacteria. This was evident by the high amount of acetate removed with minimal reduction in sulphate. There is a need for the development of a more robust and versatile SRB community with sufficient colonisation of the secondary reactor prior to dual operation. The pre-colonisation of the secondary reactor before dual operation to establish an active SRB population may facilitate enhanced treatment of the residual sulphate from the primary reactor.

The study highlighted the temperature dependence and limitations of all three reactors at low temperature. Despite having been exposed to 10°C, low levels of sulphate reduction were still maintained and was not completely inhibited. In addition, all three reactors were able to recover sulphate conversion once favourable operating conditions were restored. The ability to recover performance highlighted the resilience of the hybrid process to overcome exposure to stress conditions experienced at low temperature. This is largely linked to the versatility and robustness of the microbial communities responsible for the process reactions.

Chapter 9

Effect of feed sulphate concentration

9.1 Introduction

Biological treatment of sulphate-rich waste streams is dependent on the initial feed sulphate concentration as well as its loading rate. In these processes the sulphate loading rate can be mediated by dilution rate (HRT) or the feed sulphate concentration. In Chapters 4, 5 and 6 the effect of HRT on the performance of the hybrid LFCR process was evaluated. These studies proved effective in establishing an optimal HRT at 2 days based on sulphate reduction, sulphide oxidation and sulphur recovery. Furthermore, the studies assessed the effect of reactor geometry as well as the use of different electron donor on process performance under changing HRT conditions.

In a wastewater environment, the initial sulphate concentration can vary and may change relative the conditions and source of the waste stream. Sulphate-rich contaminated wastewater can range between 1 - 10 g/L and can have a significant impact on process performance (Brahmacharimayum et al., 2019). Several studies have evaluated the effect of sulphate concentration on biological sulphate reduction under different reactor configurations and operating parameters (Erasmus, 2000; Moosa et al., 2002; Al-zuhair et al., 2008; Oyekola et al., 2012). In the current work the hybrid LFCR represents a unique system in which both sulphate reduction and sulphide oxidation occurs simultaneously. The effects of feed sulphate concentration can affect both biological processes; these have not been evaluated in the hybrid LFCR system.

Therefore, an investigation into the effects of sulphate loading on the performance of the hybrid LFCR was critical to further characterise the process. In this study the assessment of feed sulphate concentration on process performance was performed. The study evaluates the potential of the process to effectively treat and maintain process performance over a range of applied feed sulphate concentrations.

The main objectives addressed in this chapter are as follows:

1. Evaluate the effect of feed sulphate concentration on biological sulphate reduction, sulphide oxidation and sulphur recovery performance.
2. Investigate the effects of sulphate loading on process performance as a function of reactor geometry and scale within the 2 L and 8 L lactate-fed reactors.
3. Assess the performance of a dual 8 L lactate-fed reactor system for enhanced process efficiency.

The chapter is structured as a study into the effect of feed sulphate concentration on process performance within the lactate-fed reactors, followed by an investigation into the dual operation of the 8 L lactate-fed reactor.

9.2 Effect of sulphate concentration loading

9.2.1 Experimental approach

After evaluating the effects of temperature in Chapter 8, the reactors were placed on a 4 day HRT at 25°C to recovery process performance. In this study an investigation into the effect of feed sulphate concentration on the 2 L and 8 L lactate-fed reactor was performed. The study evaluated a range of feed sulphate concentrations (1 g/L, 2.5 g/L, 5 g/L, and 10 g/L) operated at ambient temperature and at a 4 day HRT. The feed COD/SO_4 ratio was maintained at 0.7 with the feed lactate concentration increased accordingly. This ensured that any observed trend was a function of the overall impact of varying feed sulphate concentration and not due to carbon limitation. For each sulphate loading condition, the system was operated continuously during which biofilm disruption was performed intermittently followed by a biofilm harvest at the end of each experimental run (feed sulphate concentration adjustment). The reactor performance was monitored regularly (every second day) as described in Section 3.2. The harvested FSB was collected, dried and stored for elemental analysis. System performance was evaluated based on sulphate conversion, volumetric sulphate reduction rate, volumetric sulphide oxidation rate, sulphide conversion and elemental sulphur recovery.

9.2.2 Results and Discussion

9.2.2.1 Effect of Sulphate loading on reactor performance

The reactor performance data as a function of feed sulphate concentration is shown in Figure 9.1 to Figure **9.12**. During initial operation at 1g/L feed sulphate concentration, the measured residual sulphate concentration (Figure 9.1) was relatively stable in both 2 L and 8 L lactate-fed reactors. In the 2 L lactate-fed reactor (Figure 9.1 A), variation in residual sulphate concentration between reactor samples (FM, FB, BM, and BB) was observed. The largest deviation occurred within the FB sample which had lower residual sulphate concentrations compared to the average across the other reactor samples. This was similarly observed in Section 8.3.2.1, which was attributed to the high accumulation of biomass that extended above the FB sampling port. As a result, high concentration of biomass was observed within the FB sample and was accompanied by a low residual sulphate concentration. In contrast, the residual sulphate concentration measured within the 8 L lactate-fed reactor was consistent and relatively stable with minimal difference observed between reactor samples. This was consistent with the finding in Section 8.3.2.1 where the degree of biomass accumulation near the FB port was not as extensive and did not affect the FB sample as within the 2 L reactor.

After increasing the feed sulphate concentration to 2.5 g/L there was an initial sharp increase in residual sulphate concentration, followed by a gradual decrease as the systems acclimatised and recovered (Figure 9.1). This was expected as the microbial community transition and adaptation to the increase in sulphate and lactate concentration occurred. By day 51 the biofilms were harvested and the reactors were operated for an additional period, up until day 88, in order to stimulate higher sulphate conversion. This proved an effective strategy as the residual sulphate concentration steadily decreased in both the 2 L (19.0 to 10.8 mmol/L) and 8 L (20.6 to 14.6 mmol/L) reactors, achieving a sulphate conversion of 59 and 44%, respectively.

Interestingly, towards the end of the experimental run at 2.5 g/L feed sulphate concentration, the frequent sampling of the FB port over time had resulted in a decrease in the biomass concentration at the site of sampling. Eventually the FB sample became more representative of the bulk volume of the reactor, with a similar consistency (planktonic) to that of the other reactor samples (FM, BM and BB). As a result, less deviation in residual sulphate concentration was observed within the FB sample, supporting the role of biomass in causing the variation between bulk samples.

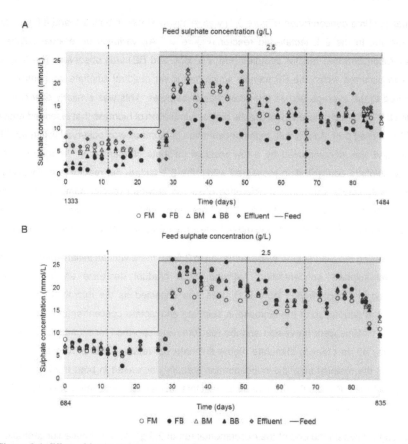

Figure 9.1: Effect of feed sulphate concentration at 1 and 2.5 g/L on residual sulphate concentration as a function of time across reactor sampling ports (FM, FB, BM, BB) and effluent in the A) 2 L lactate-fed and B) 8 L lactate-fed reactors. Biofilm disruption and harvest events are indicated by vertical dotted and solid lines, respectively. A change in feed sulphate concentration is indicated by the transition in shading intensity and was accompanied by a biofilm harvest.

Further increase in feed sulphate concentration to 5 and 10 g/L had a similar impact on the residual sulphate concentration (Figure 9.2). Initially, there was an observed increase the measured residual sulphate concentration before gradually decreasing over time. Noticeably, lower sulphate concentrations were consistently measured within the 2 L reactor in comparison with the 8 L reactor. Furthermore, the intermittent disruption and harvesting of the biofilm had minimal impact on the residual sulphate concentration. This was largely attributed to the extensive biomass accumulation within both reactors, which enhanced process stability over time such that the ingress of oxygen after disruption or harvesting of the biofilm had less of an effect on the SRB population.

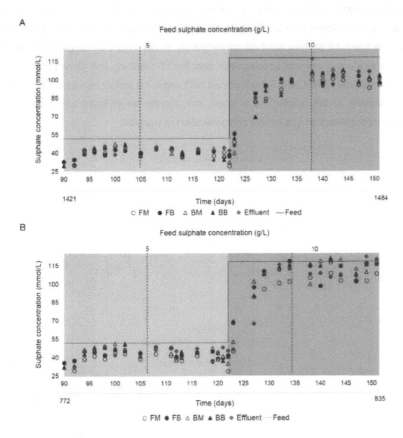

Figure 9.2: Effect of feed sulphate concentration at 5 and 10 g/L on residual sulphate concentration as a function time across reactor sampling ports (FM, FB, BM, and BB) and effluent in the A) 2 L and B) 8 L lactate-fed reactors. Biofilm disruption and harvest events are indicated by vertical dotted and solid lines, respectively. A change in feed sulphate concentration is indicated by the transition in shading intensity and was accompanied by a biofilm harvest.

The dissolved sulphide concentration measured over the duration of the experimental study is shown in Figure 9.3. A similar profile in response to increasing feed sulphate concentration from 1 to 5 g/L was observed across both 2 L and 8 L reactors. Notably, there was an increasing trend in the maximum sulphide concentration as the feed sulphate concentration was increased from 1 to 5 g/L in the 2 L (4.7 to 16.0 mmol/L) and 8 L (3.9 to 15.0 mmol/L) reactors, respectively. In addition, the rate at which the sulphide increases after biofilm disruption or harvest became more pronounced as the feed sulphate was increase. These results corresponded well with the residual sulphate data, shown in Figure 9.1 and Figure **9.2**.

In the 2 L lactate-fed reactor there was some variation observed amongst the reactor sampling ports, where higher sulphide concentration was measured in the FB samples. This was consistent with the lower residual sulphate concentration measured in the FB sample while operated at 1 and 2.5 g/L feed sulphate concentration. It was previously suggested in Section 8.3, that the extent of biomass accumulation near the FB sampling port formed a differential gradient across the biofilm, where the localised HRT within the biofilm was longer than that of the bulk volume of the reactor. As a result, the discrete environment within the biofilm facilitated high conversion of sulphate and generation of sulphide.

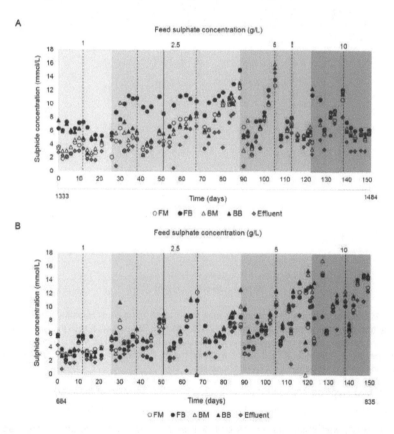

Figure 9.3: Effect of feed sulphate concentration on dissolved sulphide concentration as function of time across reactor sampling ports (FM, FB, BM, and BB) and effluent in the A) 2 L and B) 8 L lactate-fed reactors. Biofilm disruption and harvest events are indicated by vertical dotted and solid lines, respectively. A change in feed sulphate concentration is indicated by the transition in shading intensity and was accompanied by a biofilm harvest.

As expected, the effluent sulphide concentration was consistently lower than the bulk volume of the reactor. During operation at 5 g/L feed sulphate concentration on day 113 a premature biofilm disruption in the 2 L reactor caused a rapid decrease in dissolved sulphide concentration.

The pH data across both 2 L and 8 L reactors over the duration of the experimental study is shown in Figure 9.4. Most notably, there was an increase in the measured pH as the feed sulphate concentration increased from 1 to 5 g/L in both reactors. A consequence of increased alkalinity generation through sulphate reduction. In addition, the release of hydroxyl ions during partial sulphide oxidation to elemental sulphur was responsible for the elevated effluent pH.

At a feed sulphate concentration of 10 g/L the pH measured in the 2 L reactor initially decreased before recovering a pH of 8.5. In comparison a rapid decrease in pH occurred within the 8 L reactor, which ranged between 7.1 and 7.3. These results coincided with the observed decrease in sulphate reduction within the 8 L reactor at 10 g/L. In sulphidogenic reactors the pH plays a critical role in the inhibition of sulphide to SRB activity (Moosa et al., 2006; van den Brand et al., 2016). Sulphide can be present with in liquid in various states of which undissociated H_2S has the strongest inhibitory effect due to its ability to permeate the cell membrane and resulting in denaturation of enzymes. The quantity of $H_2S_{(g)}$ is largely determined by the pH where hydrogen sulphide exists as a mixture of $H_2S_{(g)}$ and HS^- between pH 6 to 8 (Moosa & Harrison, 2006). Below pH 6, undissociated $H_2S_{(g)}$ dominates while at a pH >7.5, the $H_2S_{(g)}$ fraction of the total sulphide in solution is minimal (van den Brand et al., 2016). Previous research has shown that biofilm systems (biomass retention) can withstand higher levels of sulphide (up to 6.2 mmol-H_2S/L and 31.2 mmol-DS/L), while suspended growth systems (planktonic) showed greater sensitivity to much lower levels (1.9 mmol-H_2S/L and 4.7 mmol-DS/L) (Maillacheruvu et al., 1993). The higher tolerance of biofilms to sulphide was attributed to the diffusion gradient across the biofilm, where sulphide concentration decreases with decreasing biofilm depth protecting the cells from exposure to inhibitory concentrations of sulphide. Similarly, In the current study, the extent of biomass accumulation within the 2 L reactor compared to the 8 L reactor may have conferred higher system robustness. Though there was a significant accumulation of the biofilm within the 8 L reactor and colonisation on the carbon microfibers, the bulk volume of the reactor was largely planktonic and susceptible to change in operating conditions. Considering these findings, the decrease in sulphate reduction within the 8 L reactor during operation at 10 g/L was likely a consequence of sulphide inhibition, evident by the concomitant decrease in pH below 7.5 and high sulphide concentration (14.9 mmol/L) in the reactor samples.

Overall, the pH in both reactors was consistently higher than the feed (pH 7.0) which demonstrated the increased buffering capacity of the system as the feed sulphate concentration was increased, largely attributed to the increase in BSR and generation of bicarbonate. The high pH >7.0 insured that the sulphide generated remained in solution as HS⁻ with minimal liberation of undissociated sulphide ($H_2S_{(g)}$).

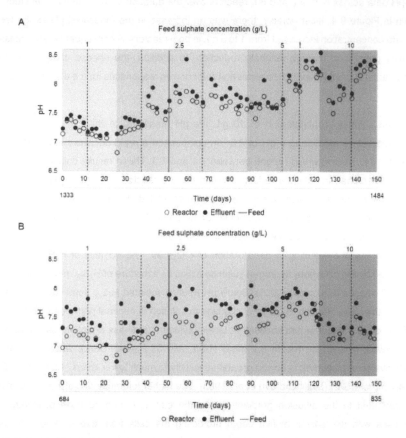

Figure 9.4: Effect of feed sulphate concentration on mean pH data of reactor samples (FM, FB, BM, and BB) and effluent as a function of time in the A) 2 L and B) 8 L lactate-fed reactors. Biofilm disruption and harvest events are indicated by vertical dotted and solid lines, respectively. A change in feed sulphate concentration is indicated by the transition in shading intensity and was accompanied by a biofilm harvest.

The VFA concentration profile analysis for both 2 L and 8 L lactate-fed reactors are shown in Figure 9.5 and Figure 9.6. Consistent with previous investigations, complete utilisation of lactate was observed at 1 g/L feed sulphate concentrations in both 2 L and 8 L reactors. An increase in the residual lactate concentration at 2.5 g/L feed sulphate concentration, occurred as the feed concentration was adjusted accordingly to maintain a 0.7 COD/SO$_4$ ratio. Throughout the study the acetate and propionate concentration remained below 18 mmol/L regardless of the increased lactate concentration (Figure 9.6).

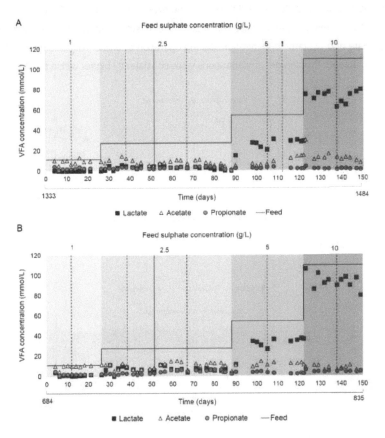

Figure 9.5: Effect of feed sulphate concentration on the volatile fatty acid profile as a function of time in the A) 2 L and B) 8 L lactate-fed reactors. Data represents mean values obtained from the reactor sampling ports (FM, FB, BM, and BB). Biofilm disruption and harvest events are indicated by vertical dotted and solid lines, respectively. A change in feed sulphate concentration is indicated by the transition in shading intensity and was accompanied by a biofilm harvest.

Shortly after increasing the feed sulphate concentration to 2.5 g/L there was a pronounced increase in residual lactate concentration accompanied by a sharp decrease in acetate concentration within both reactors (Figure 9.6). Furthermore, the marked increase in propionate, particularly in the 8 L reactor, suggested that an increased proportion of lactate was directed toward fermentation. The increase in lactate concentration which accompanied the increase in sulphate loading favoured the growth of fermentative microorganisms. These microorganisms are characterised by a higher K_s and μ_{max}, which were able to proliferate due to the increased lactate availability at the higher sulphate loading (Oyekola et al., 2010). Over time as the system acclimatised to a 2.5 g/L feed sulphate concentration, the concentration of lactate decreased gradually. Toward the end of the experimental run at 2.5 g/L feed sulphate concentration, in the 2 L reactor there was a decrease in lactate concomitant with a decrease in propionate and acetate concentration. During operation at 5 and 10 g/L, concentration of acetate increased. Propionate concentrations were consistently higher within the 8 L reactor.

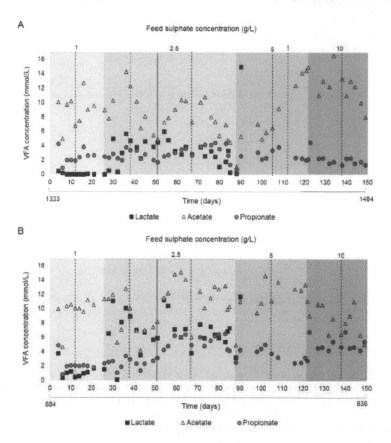

Figure 9.6: Effect of feed sulphate concentration of volatile fatty acid profile showing the main lactate metabolism acetate and propionate as a function of time in the A) 2 L and B) 8 L lactate-fed reactors. Data represents mean values from reactor sampling ports (FM, FB, BM, and BB). Biofilm disruption and harvest events are indicated by vertical dotted and solid lines, respectively. A change in feed sulphate concentration is indicated by transitioning shading intensity and was accompanied by a biofilm harvest.

Analysis of the volumetric rates associated with lactate utilisation and production of acetate and propionate are shown in Figure 9.7. The results reveal a sharp increase in the volumetric lactate utilisation rate (VLUR) when the feed sulphate concentration was increase from 1 to 2.5 g/L in both reactors. In the 2 L reactor the increase in VLUR was accompanied by an increase in VSRR and a decrease in the volumetric acetate production rate.

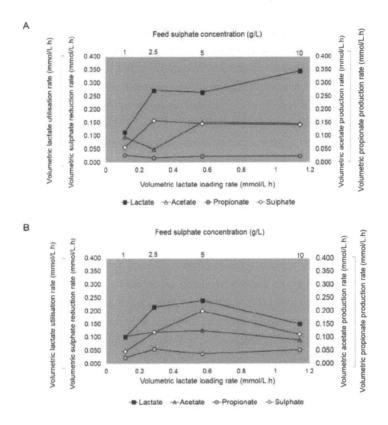

Figure 9.7: Effect of feed sulphate concentration on lactate utilisation, acetate and propionate production and sulphate reduction rates within the A) 2 L and B) 8 L lactate-fed reactors.

The volumetric propionate production rate remained relatively constant in both 2 L and 8 L reactors over the duration of the study. On average, higher propionate production rates were recorded within the 8 L compared to the 2 L reactor. The highest propionate production rate was observed in the 8 L reactor while operated at a 2.5 g/L feed sulphate concentration. Minimal change in the VLUR was observed in both 2 L and 8 L reactors when the feed sulphate concentration was increased from 2.5 to 5 g/L. This was accompanied by a slight decrease in VSRR within the 2 L reactor and an increase in VSRR in the 8 L reactor at a 5 g/L feed sulphate concentration.

Upon further increased to a 10 g/L feed sulphate concentration, the reactors exhibited different results with an increase in the 2 L reactor and decrease in the 8 L reactor. The decrease in VLUR exhibited in the 8 L reactor coincided with the decrease in VSRR. In contrast, there was minimal change in the VSRR within the 2 L reactor while operated at a 10 g/L feed sulphate concentration. These results could reflect the microbial population and environmental conditions within the different reactors. As previously discussed there was a marked decrease in pH (Figure 9.4 B) within the 8 L reactor during operation at 10 g/L which may have resulted in the exposure to inhibitory concentration of undissociated H_2S. Alternatively, the difference in performance could be a function of the microbial population in which the extent of biomass within the 2 L reactor provided a more robust SRB community. This was highlighted by the higher VLUR within the 2 L reactor which indicated that the reactor was more active and less affected by the increase in feed sulphate concentration.

From these results it became clear that lactate metabolism toward sulphate reduction was favoured in both reactors throughout the study. This was largely due to the operation of the reactors at a long HRT of 4 days, which favoured SRB activity. It is important to re-emphasise that lactate was supplemented at a COD/SO_4 ratio of 0.7, which is slightly higher than the theoretical ratio (0.67) to achieve 100% sulphate conversion through complete oxidation. However, in the current study, sulphate reduction predominantly occurred via incomplete oxidation to acetate of which only 53% sulphate conversion was theoretically possible. Taking this into consideration the sulphate reduction performance in the reactors was limited via partial oxidation of the substrate (Reaction 9.2), with the complete oxidation of lactate being the rate limiting step in the current investigation (Reaction 9.1).

9.2.2.2 Stoichiometric dependency on sulphate loading

The theoretical (Table 9.1) and experimental (Table 2) stoichiometric ratio profiles as a function of feed sulphate concentration are presented in Figure 9.8. The experimental stoichiometric ratio L:S correspond largely with the theoretical values of incomplete lactate oxidation via Reaction 9.2. The L:S and A:S ratios < 2 suggests that complete lactate oxidation likely occurred in both 2 L and 8 L reactors (Table 9.2). It is likely that an active acetate utilising microbial community including SRB and non-SRB were present within the reactors.

Table 9.1: Theoretical ratios of the metabolic reactions associated with the respective electron donors.

Reaction No.	Chemical reaction	Theoretical stoichiometric ratio		
		L:A	L:S	A:S
9.1	$2\ Lactate + 3\ SO_4^{2-} \rightarrow 6\ HCO_3^- + 3\ HS^- + H^+$	-	0.67	-
9.2	$2\ Lactate^- + SO_4^{2+} \rightarrow HS^- + 2\ Acetate^- + 2\ HCO_3^- + H^+$	1.0	2.0	2.0
9.3	$3\ Lactate \rightarrow acetate + 2\ propionate + HCO_3^- + H^+$	3.0	-	-

The high L:A ratio at 2.5 g/L feed sulphate concentration in the 2 L reactor was attributed to the low residual acetate concentration (Figure 9.5 A). Considering that the lactate was completely utilised it was expected that a large proportion of acetate would be generated based on incomplete oxidation (Reaction 9.2). Instead the low acetate measured, implied that acetate utilisation by SRB and non-SRB was favoured. This finding was further strengthened by the high L:A ratio concomitant with the low A:S ratio. While the L:A ratio (>1) indicated that less acetate was produced than the amount of lactate utilised, the low A:S (<2) indicated that less acetate was produced than the amount of sulphate reduced via incomplete oxidation (Reaction 9.2). Together these ratios suggest that complete oxidation (Reaction 9.1) was present. The deviation of the experimental ratio L:S at 5 and 10 g/L within the 8 L reactor was below the theoretical value for both complete or incomplete lactate oxidation, which indicated that less lactate was utilised than the theoretical requirement for the exhibited sulphate reduction.

In addition, with the estimated experimental ratios, the carbon balance ratio calculated based on the total moles C measured and the total moles C of lactate fed into the system was determined (Table 9.2). Values ranging from 0.2 to 1 was exhibited in both reactors over the duration of the study. The low values that deviated from the carbon balance < 1 indicated that a portion of the available lactate was completely metabolised to CO_2. These values agreed with the experimental ratios (L:A, L:S and A:S) that a portion of the lactate was completely oxidised to CO_2 and therefore would not be accounted for in the overall carbon balance.

Table 9.2: Dependency of molar ratio of lactate utilised to the other substrates involved in biological sulphate reduction on feed sulphate concentration, using lactate as the sole carbon-source and electron donor. Average values of experimental stoichiometric ratios are compared with the theoretical ratios (Table 9.1). The carbon balance of total moles VFA accounted compared with the total amount of lactate fed is also presented.

Feed sulphate concentration (g/L)	Volumetric rates (mmol/L.h)				Stoichiometric ratios [a]			Carbon balance
	Lactate utilisation rate	Acetate production Rate	Propionate production rate	Sulphate reduction rate	Total moles lactate used/mole acetate produced (L:A)	Total moles lactate used/mole sulphate reduced (L:S)	Total moles acetate produced/ mole sulphate reduced (A:S)	Total C moles out/ Total C mole lactate fed [b] (Effluent:Influent)
2 L lactate-fed								
1	0.114	0.096	0.027	0.056	0.9	1.3	1.5	0.8
2.5	0.272	0.049	0.016	0.159	6.5	1.6	0.3	0.2
5	0.264	0.151	0.023	0.146	1.6	1.5	1.0	0.8
10	0.345	0.147	0.024	0.142	2.3	2.2	1.0	0.8
8 L lactate-fed								
1	0.100	0.124	0.023	0.046	0.6	1.4	2.3	0.9
2.5	0.215	0.120	0.055	0.119	1.9	1.5	0.8	0.7
5	0.240	0.125	0.037	0.200	1.4	0.8	0.6	0.8
10	0.195	0.126	0.053	0.110	2.1	1.0	0.4	1.0

[a] Calculated, excluding the contribution of fermentation based on residual propionate concentration
[b] Carbon balance of the total mol VFA measured to the total amount of mol lactate-fed

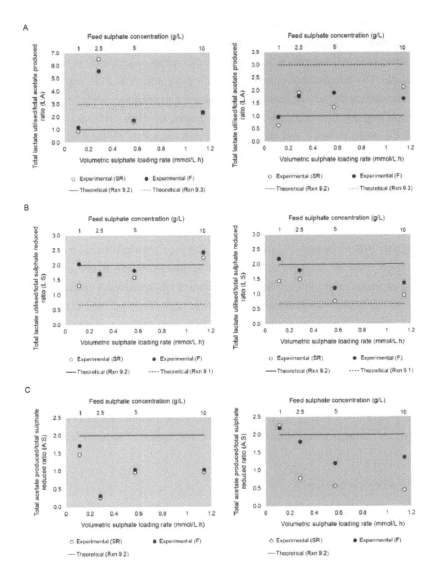

Figure 9.8: Experimental data of the hybrid LFCR reactors investigating the effect of feed sulphate concentration (1.0 to 10 g/L) at 4 day HRT (dilution rate:) on biological sulphate reduction stoichiometry in the 2 L (left) and 8 L (right) reactors. A) Total moles of lactate utilised per mole total acetate produced (L:A), B) Total moles of lactate utilised per total moles sulphate reduced (L:S), C) Moles of acetate produced per total moles of sulphate reduced (A:S). Experimental ratio with (F) and without (SR) the contribution of fermentation, calculated stoichiometrically (Rxn 9.3) based on residual propionate concentration. The horizontal solid (Rxn 9.2) and dotted (Rxn 9.1 and 9.3) lines represent the theoretical ratio for the respective reactions.

Although lactate was completely utilised at 1 and 2.5 g/L feed sulphate concentration, at 5 and 10 g/L the utilisation decreased. Similarly, the high L:A ratio together with the low effluent: influent carbon ratio (<1) suggests that complete oxidation was present even at high lactate concentrations. This reflects the diverse microbial community within the LFCR. In a well-mixed CSTR, under controlled operating conditions, the system will favour and select for the growth of incomplete oxidising SRB and will washout slow growing complete oxidisers over time. In the LFCR the ability to retain and accumulate biomass promoted the establishment of a complete oxidising community. Although their activity and growth are relatively low, over an extended period of operation, these communities are able to develop within the reactor. Similarly, the consistent production of propionate observed throughout the study even at a long HRT of 4 days, highlight the activity of a fermentative microbial community within the LFCR. Though their activity is not favoured under longer HRT operating conditions when competing with SRB, fermentative microorganism is able to persist within the reactor attaching to the support material, incorporated within biofilms and operating at low activity until favourable condition arise. The preferential attachment of fermentative microorganisms belonging to the phylum *Bacteroidetes, Synergistetes and Firmicutes* was highlighted in Chapter 7. Though fermentative microorganism were not dominant within the reactor compared to SRB genera they were present at relatively high abundance and were able to assimilate a consistent amount of lactate with an average propionate production rate of 0.023 and 0.042 mmol/L.h in the 2 L and 8 L reactors, respectively. Ultimately when considering that lactate was supplemented near the theoretical COD/sulphate ratio (0.7) for complete sulphate conversion, the degree of lactate metabolism toward fermentation can have major implications on overall process performance, limiting the amount of substrate available for BSR.

9.2.2.3 Sulphate reduction kinetics

The kinetics of anaerobic sulphate reduction at various feed sulphate concentrations of 1, 2.5, 5 and 10 g/L in both 2 L and 8 L reactors are shown in Figure 9.9. The results revealed an increase in VSRR as the feed sulphate concentration increased to 2.5 g/L which was accompanied by an increase in the volumetric lactate utilisation rate (VLUR). The increase in sulphate loading rate mediated through increasing the feed sulphate concentration had a positive impact on the VSRR, however this was accompanied by a decrease in sulphate conversion. A similar impact on sulphate reduction performance was observed during the effect of HRT studies reported in Section 5.4.2.3 and 6.3.2.3.

Upon further increase to 5 and 10 g/L feed sulphate concentration, the 2 L and 8 L reactors exhibited distinctly different sulphate reduction kinetics. In the 2 L reactor the VSRR remained relatively constant when increasing the feed sulphate concentration from 2.5 to 10 g/L (Figure 9.9 A). In contrast, the 8 L reactor exhibited a linear increase in VSRR from 1 to 5 g/L before decreasing at a 10 g/L feed sulphate concentration (Figure 9.9 B). The ability of the 2 L reactor to maintain BSR even at the high feed sulphate concentration of 10 g/L was largely attributed to the extent of biomass accumulation within the reactor. In the 8 L reactor the marked decrease in VSRR at 10 g/L feed sulphate concentration coincided with the decrease in pH (Figure 9.4). The inhibition at high sulphide concentration was discussed in Section 9.2.2.1.

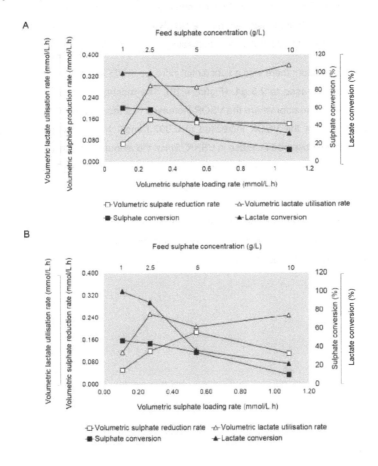

Figure 9.9: Effect of feed sulphate concentration on biological sulphate reduction kinetics showing the volumetric rates and conversion of sulphate reduction and lactate utilisation in the A) 2 L and B) 8 L lactate-fed reactors.

9.2.2.4 Sulphide oxidation kinetics

The sulphide oxidation performance in the lactate-fed reactors as a function of feed sulphate concentration over time is shown in Figure 9.10. In both 2 L and 8 L reactors, at a 1 g/L feed sulphate concentration, the VSOR increased rapidly after a biofilm disruption event, reaching a maximum of 0.050 and 0.028 mmol/L.h, respectively. As the biofilm regenerated at the surface and impeded oxygen penetration into the bulk volume, the VSOR decreased to a minimum. The distinctive trend in VSOR was consistent with results obtained in Section 5.4.2.4 and 8.3.2.5, which demonstrated that the VSOR within the hybrid LFCR process was largely mediated through the disruption and harvesting of the biofilm.

In both 2 L and 8 L reactors there was a substantial increase in VSOR once the feed sulphate concentration was increased to 2.5 g/L (Figure 9.10). The most pronounced increase was observed within the 2 L reactor where the VSOR increased to a maximum of 0.107 mmol/L.h, while in the 8 L reactor the VSOR increased to 0.097 mmol/L.h. The increase in VSOR corresponded with the observed increase in VSRR. Since the amount of sulphide in the hybrid LFCR is dictated by the rate of sulphate reduction, the VSRR effectively represents the sulphide loading rate in the system for partial sulphide oxidation. Therefore, the increase in VSRR (sulphide loading) resulted in an increase in VSOR. Previous study by Dogan et al. (2012) demonstrated biological sulphide oxidation in an airlift reactor under oxygen limiting conditions. The study revealed that the increase in volumetric sulphide loading rate resulted in an increase in sulphide oxidation and elemental sulphur production.

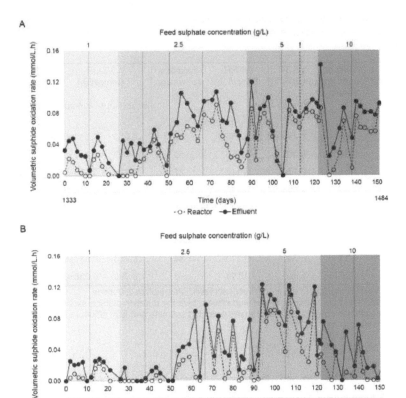

Figure 9.10: Effect of feed sulphate concentration on volumetric sulphide oxidation rate and conversion over time in the A) 2 L and B) 8 L lactate-fed reactors. Biofilm disruption and harvest events are indicated by vertical dotted and solid lines, respectively. A change in feed sulphate concentration is indicated by the transition in shading intensity and was accompanied by a biofilm harvest.

Although a high VSOR was achieved shortly after a biofilm disruption and harvest event, it could not be maintained and sharply decrease to a minimum. Considering that the average VSOR (Table 9.3) observed over the duration of each experimental run was lower than the maximum, the cumulative sulphide conversion ranged between 31 and 59% across both reactors. The potential sulphide conversion obtainable within the hybrid LFCR was limited by the infrequent disruption regime adopted within this study (±16 days). The decision to limit the rate of biofilm disruption was to ensure sulphate reduction activity was not compromised as this would have severe implications on the overall process performance as previously shown in Section 5.4.2.

Table 9.3: Overall process performance as a function of increasing feed sulphate concentration showing the sulphate reduction and sulphide oxidation kinetics within the 2 L and 8 L lactate-fed reactors.

Sulphate concentration (g/L)	Period of operation (days)	VSLR (mmol/L.h)	VSRR (mmol/L.h)	Sulphate conversion (%)	VSOR (mmol/L.h)		Sulphide conversion (%) [c]
					Maximum [a]	Average [b]	
2 L LFCR (Lactate-fed)							
1	26	0.146	0.056	52	0.050	0.030±0.03	49
2.5	37	0.365	0.159	59	0.107	0.075±0.03	59
5	34	0.730	0.146	27	0.129	0.058±0.02	42
10	29	1.459	0.142	13	0.142	0.076±0.02	46
8 L LFCR (Lactate-fed)							
1	26	0.146	0.046	42	0.028	0.016±0.03	38
2.5	37	0.365	0.119	44	0.097	0.045±0.02	52
5	34	0.730	0.200	34	0.115	0.037±0.01	39
10	29	1.459	0.110	10	0.098	0.038±0.01	31

[a] Maximum VSOR recorded in the final effluent achieved after biofilm disruption
[b] Average VSOR measured based on the final effluent over the duration of each experimental run
[c] Cumulative sulphide conversion based on the expected sulphide and final effluent over the duration of each experimental run

9.2.2.5 Biofilm harvest and sulphur recovery

The total amount of harvested FSB recovered at the end of each experimental run is shown in Figure 9.11. A substantial increase in the amount of recovered FSB in both 2 L (2.4 to 5.7 g) and 8 L (10.6 to 19.6 g) reactors were observed when the sulphate loading rate was increased from a feed sulphate concentration of 1 g/L to 2.5 g/L. A further increase in feed sulphate concentration to 10 g/L resulted in a decrease in the total amount of biofilm harvested. The increase in the amount FSB recovery coincided with the observed increase in VSRR and VSOR (Figure 9.9; Table 9.3).

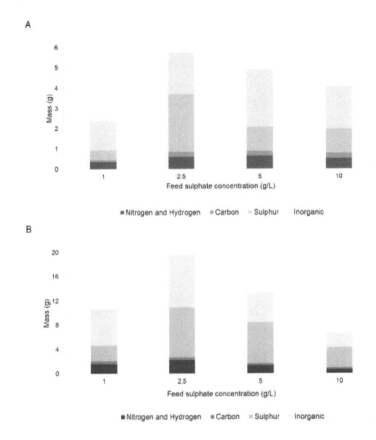

Figure 9.11: Effect of feed sulphate concentration on the recovery of floating sulphur biofilm, showing the elemental composition of nitrogen and hydrogen, carbon, sulphur and inorganics based on the amount (grams) of biofilm harvested in the A) 2 L and B) 8 L lactate-fed reactors.

Elemental composition (%) analysis (Figure 9.12) of the FSB showed a marked increase in sulphur content in both 2 L (20 to 50%) and 8 L (24 to 42%) reactors after increasing the feed sulphate concentration to 2.5 g/L. Upon further increase in the feed sulphate concentration to 5 g/L and 10 g/L, there was a decrease in sulphur content in the 2 L reactor. In contrast, the FSB recovered from the 8 L reactor gradually increased in sulphur composition. Interestingly, the inorganic fraction which largely made up the FSB during operation at 1 g/L feed sulphate concentration decreased when changed to 2.5 g/L in both 2 L (60 to 35%) and 8 L (56 to 44%) reactors.

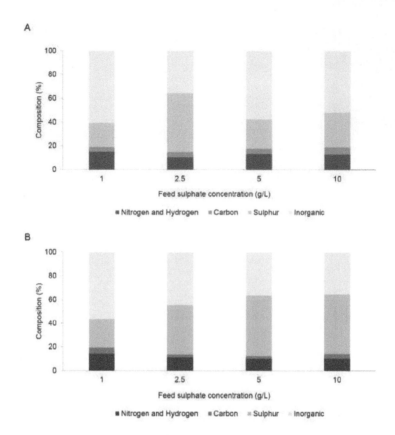

Figure 9.12: Effect of feed sulphate concentration on the elemental composition of the floating sulphur biofilm, showing the composition (%) of inorganics, sulphur, carbon, nitrogen and hydrogen present in the harvested FSB collected from the respective experimental runs in the A) 2 L and B) 8 L lactate-fed reactors.

Sulphur recovery through harvesting of the FSB was evaluated based on three performance variables: 1) the total sulphate-S load during the operation at each feed sulphate concentration; 2) the expected sulphide-S (sulphate reduced); and 3) the converted sulphide-S; where the values are expressed as a total contribution in grams sulphur, denoted by (-S). Based on the total sulphate-S load, the elemental sulphur recovery through the FSB ranged between 2 and 18% across both reactors. However, when considering the sulphur recovery based on the expected and converted sulphide-S parameters, there was a significant increase (Table 9.4). While the sulphur recovery based on the expected sulphide-S ranged between 15 and 32%, the converted sulphide-S ranged between 36 and 85% across both 2 L and 8 L reactors.

The minimal increase in sulphate concentration within the effluent (Figure 9.1 and Figure 9.2) confirmed that there was negligible complete oxidation of the sulphide to sulphate. As a result it is possible that the large proportion of sulphur, represented as gap-S, that was not recovered from the FSB was predominantly in the form of fine colloidal sulphur particles and fragments of biofilm that were dispersed into the bulk volume and released into the effluent stream. The dissolution and accumulation of elemental sulphur into the effluent stream, was evaluated and confirmed in Section 9.3.2.

Overall evaluation of the sulphur recovery through the FSB revealed that the major limitations of the hybrid LFCR was largely attributed to the sulphate reduction performance and regulation of the biofilm disruption and harvesting regime. While the increase in HRT may be a potential solution to achieve higher sulphate conversion it comes at a compromise of the VSRR. One strategy that was evaluated for overcoming the low sulphate conversion and sulphur recovery within the hybrid LFCR was through the operation of an additional LFCR unit downstream as a dual reactor system. This was explored in Section 8.4 in the 2 L acetate-fed reactor during which the additional recovery of FSB within the secondary reactor increased the overall sulphur recovery.

Table 9.4: Summary of the sulphur recovery in the 2 L and 8 L lactate-fed reactors as a function of feed sulphate concentration showing the amount of sulphur recovered from harvesting of the biofilm (FSB-S) as well as the respective recovery (%) based on each parameter including the total sulphate-S, expected sulphide-S (sulphate reduced: SR) and converted sulphide-S (SO).

Sulphate concentration (g/L)	FSB-S harvested (g)	Total sulphate-S [a]		Expected sulphide-S				Converted sulphide-S			
		Amount (g)	Recovery (%) [b]	SR (%)	Amount (g)	Recovery (%) [b]	SO (%)	Amount (g)	Recovery (%) [b]	Gap-S (g) [c]
				2 L lactate-fed							
1	0.5	4.6	11	52	2.4	20	49	1.2	42	0.7
2.5	2.9	16.4	18	59	9.7	30	59	5.7	50	2.8
5	1.2	30.1	4	27	8.1	15	42	3.4	36	2.2
10	1.2	51.4	2	13	6.7	18	46	3.1	39	1.9
				8 L Lactate-fed							
1	2.6	19.1	14	42	8.0	32	38	3.1	85	0.5
2.5	8.2	67.8	12	44	29.9	28	52	15.4	54	3.6
5	6.8	124.7	6	34	42.4	16	39	16.5	41	6.7
10	3.3	212.7	2	10	21.3	16	31	6.5	51	1.8

[a] Total sulphur load over the duration of the experimental run
[b] Recovery based on FSB-S
[c] Sulphur fraction of converted sulphide not recovered through the FSB-S

9.3 Dual reactor operation in series

In this section a two-stage operation of the 8 L lactate-fed reactor within a dual reactor system connected to an identical LFCR unit in series was investigated. As previously demonstrated in Section 8.4, the operation of the dual system is highly beneficial for achieving enhanced sulphur recovery in the hybrid LFCR process. Results obtained in Section 9.2.2.4 highlighted the inefficiencies in the hybrid LFCR associated with management of the biofilm. It is envisaged that the operation of the dual reactor will increase overall process performance during which key objectives to enhance sulphate conversion and sulphur recovery will be explored.

Prior to operation of the dual reactor system, the second reactor was modified with a weir at the effluent port to enhance reactor performance. As a result, a range of fluid dynamic studies was initially performed to evaluate the effect of the modification. Furthermore, the incorporation of carbon microfibers in the LFCR, which was not considered in previous dye tracer studies (Sections 5.3 and 8.2), was tested. The rationale and experimental approach pertaining to these studies will be discussed in greater detail in the following sections.

9.3.1 Experimental approach

9.3.1.1 Fluid dynamics - Modification of the LFCR

During the operation of the reactors, the effluent sulphide concentration varied between samples and resembled that of the samples derived from the ports nearest to the effluent (BM and BB). It was deduced that a portion of the bulk volume, containing high dissolved sulphide, bypassed the surface (oxidation zone) and was released into the effluent. This was identified as a possible inefficiency of the LFCR, which may have limited the maximum sulphide conversion attainable within the reactor. As a result the reactor was modified with a weir located at the effluent port (Figure 9.13). The intent was to redirect flow from the lower region of the bulk volume, near the effluent port, toward the surface. This would increase exposure of the sulphide to the oxic zone (reaction site for oxidation) before being released. It was envisaged that a lower dissolved sulphide concentration would be measured in the final effluent with less variability between effluent samples. The modification was expected to influence the hydrodynamic profile previously obtained in Section 5.3. In order to maintain consistency between reactor units, it was important to assess the impact and the significance of the modification on the overall fluid dynamics in the LFCR.

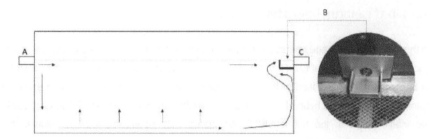

Figure 9.13: Schematic of the proposed change in mixing regime with the addition of the weir positioned at the effluent port showing A) the inlet port B) weir (modification) C) effluent port. The arrows highlight the fluid flow progression toward the effluent port the addition of the weir ensures that the bulk volume does not bypass the air-liquid interface.

The main objective of the hydrodynamic studies in this work was to verify that the LFCR design functioned optimally and conformed to the conceptual model previously described by Mooruth (2013). Although this was confirmed in previous experiments, the effect of carbon microfibers on the fluid dynamics was not considered. The addition of the carbon microfibers as a solid support for biomass attachment could potentially affect the mixing dynamics within the LFCR. Consequently, in the current investigation, a dye tracer study to determine the effects of the reactor modification (weir) and carbon microfibers was performed. The experimental set up and operation followed the procedure described in Section 3.4. The experimental runs were operated at a 2 day HRT and were performed in triplicate.

9.3.1.2 Dual operation of the 8 L lactate-fed LFCR

After assessing the hydrodynamics within the modified LFCR, a two-stage 8 L lactate-fed reactor system was set-up similarly to the dual acetate-fed system described in Chapter 8. The conceptual model of the dual reactor system is shown in Figure 8.19. The modified 8 L reactor, evaluated in the tracer experiments, served as the secondary reactor and was inoculated with overflow derived from the primary reactor. The set-up involved slightly elevating the primary reactor in order to facilitate passive flow. The dual system was operated concurrently with the effect of sulphate loading study covered in the previous section. Therefore, the results obtained from this study will demonstrate the effect of two-stage operation on process efficiency as a function of sulphate loading. The secondary reactor was sampled from the front middle (FM) and back middle (BM) sampling ports as well as the effluent. The disruption and harvesting of the FSB were performed alongside the primary reactor. However, during operation at 5 and 10 g/L feed sulphate concentration, an additional biofilm disruption, independently of the primary reactor, was performed in the secondary reactor. The objective was to demonstrate the effect of regulating the frequency of biofilm disruption on sulphide conversion and sulphur recovery.

9.3.2 Results and discussion

9.3.2.1 Fluid dynamics of the modified LFCR

The results from the dye tracer experiment with the added modification is shown in Figure 9.14. The addition of the weir at the effluent port had minimal impact on the overall mixing regime and was consistent with previous results in Sections 5.3 and 8.2 as well as the conceptual model described by Mooruth (2013). During the initial stages, upon entering the reactor, the incoming fluid sunk to the base of the reactor with a degree of back mixing occurring in the front corner (Figure 9.14 A). The fluid then proceeded along the base of the reactor toward the effluent port, while diffusive mixing was directed towards the surface and effluent port. The diffusion of the bulk volume can be observed by the change in colour intensity over time shown in Figure 9.14 B to C. The rate of diffusion from the front (inlet) to the back end (outlet) of the reactor was considerably reduced, compared to previous experiments.

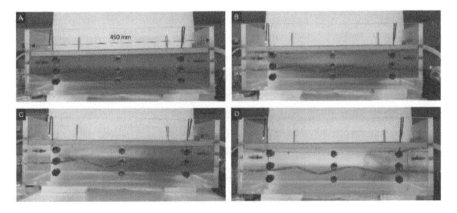

Figure 9.14: Photographic recordings showing the progression of mixing during the dye tracer experiment in the modified 8 L LFCR with added weir at the effluent port. Operated at ambient temperature and at a flow rate equivalent to a 2 day HRT. Images taken at A) 139, B) 205, C) 219 and D) 264 min, respectively.

Similarly, there was marginal impact on the overall mixing profile in the LFCR with the addition of carbon microfibers (Figure 9.15). Consistent with the previous experiment, upon entering the reactor the incoming fluid, flowed to the base of the reactor before moving along the bottom toward the effluent port. This was accompanied by diffusive mixing towards the surface and exit port over time. However, though the bulk volume had maintained the distinctive mixing pattern observed in previous experiments, there were stagnant regions that formed at the surface of the carbon microfibers, where traces of the pink dye was still present. As a result, the time taken to completely neutralise the dye (phenolphthalein) was extended.

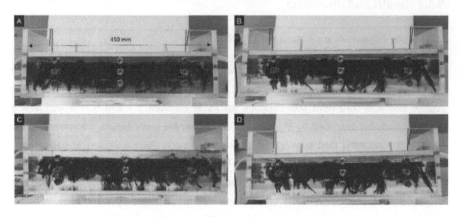

Figure 9.15: Photographic recordings showing the progression of mixing during the dye tracer experiment in the modified 8 L LFCR with added weir and carbon microfibers. Operated at ambient temperature and at a flow rate equivalent to a 2 day HRT. Images taken at A) 132, B) 204, C) 214, and D) 246 min.

The complete mixing times between the two experiments were similar, with slightly longer times recorded with the addition of the carbon microfibers (Figure 9.16). The small error bars represent the standard deviation between the replications in triplicate of each tracer study and clearly demonstrates the reproducibility of the results. In comparison to the tracer studies performed on the original reactor, a 2-fold increase in the recorded mixing time was observed. Although the complete mixing times were longer, it was still significantly shorter (approximately 10.6 times) than the operating HRT at 2 days. This indicated that the mixing within the modified LFCR, largely driven by advective and diffusive transport, was effective.

Overall, the addition of the weir and carbon microfibers had minimal impact on the hydrodynamic mixing regime within the LFCR. The reactor retained key hydrodynamic features critical for its application which included 1) the limited turbulence at the surface, 2) satisfactory mixing (largely governed by advective and diffusion transport) within the bulk volume with no observed short-circuiting and 3) displacement of the volume toward the surface and effluent. The addition of the weir proved effective and directed the bulk fluid near the effluent port to the interfacial zone before discharge through the effluent port. This aligned well with the proposed mixing profile in Figure 9.13. It is envisaged that during experimental operation, the modification (weir) will promote higher sulphide conversion by increasing the exposure of the dissolved sulphide to the oxic zone and will be evaluated in the subsequent Section 9.3.2.2 during operation of the dual reactor system.

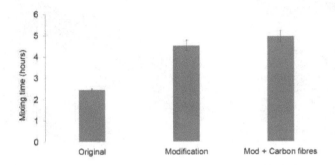

Figure 9.16: Complete mixing times comparing the original 8 L LFCR and the effects of the modification (weir) and addition of carbon microfibers. The experimental runs were performed in triplicate with the standard deviation represented by the error bars.

9.3.2.2 Effect of sulphate loading on the performance of the dual Hybrid LFCR system

The following study introduces the operation of the dual reactor system during which a comparative analysis of single-stage (Section 8.2.2) and dual operation of the 8 L reactor will be assessed to determine the impact on overall performance. After successfully assessing the hydrodynamics within the modified LFCR. The reactor was connected downstream of the 8 L lactate-fed reactor as part of a dual reactor system. Within 24 h after initial inoculation of the secondary reactor, a well-established FSB covered the entire surface of both reactors (Figure 9.17).

Figure 9.17: Dual reactor set-up of the 8 L lactate-fed LFCR showing the two identical reactors connected in series, the primary reactor (R1) was elevated above the level of the secondary reactor (R2) to ensure passive flow through into R2. The photograph was taken 21 days after biofilm disruption showing a matured FSB at the surface of both reactors.

The residual sulphate concentration profile across the dual reactor system is shown in

Figure 9.18. During operation at 1 g/L feed sulphate concentration, the residual sulphate concentration was relatively stable with minimal difference observed between R1 and R2. This suggested that there was minimal sulphate conversion within the secondary reactor. A similar result was observed within the 2 L acetate-fed dual reactor system evaluated in Section 8.4. Interestingly, once the feed sulphate concentration was increased to 2.5 g/L, there was a proportion of sulphate conversion in R2 (Figure 9.18 A), evident by the reduction in residual concentration. The higher substrate availability at 2.5 g/L may have increased the SRB activity within the secondary reactor. Since the reactor relied on the colonisation via the overflow received from the primary reactor, R2 experienced an initial acclimatisation period during which biomass accumulated within the reactor.

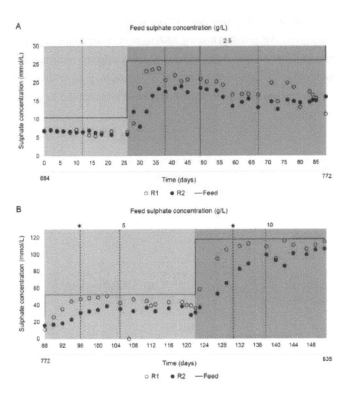

Figure 9.18: Effect of feed sulphate concentration on the performance of the dual reactor system showing the average residual sulphate concentration over time in the primary (R1) and secondary (R2) reactors A) at 1 and 2.5 g/L, as well as B) 5 and 10 g/L feed sulphate concentration. Biofilm disruption and harvest events are indicated by vertical dotted and solid lines, respectively. A change in feed sulphate concentration is indicated by the transition in shading intensity and was accompanied by a biofilm harvest. Additional biofilm disruptions conducted within R2 during operation at 5 and 10 g/L feed sulphate concentration is denoted by (★).

Once operated at 2.5 to 10 g/L feed sulphate concentration, there was an increased portion of sulphate that was reduced within the secondary reactor. In addition, a considerable amount of sulphide was converted (Figure 9.19 A) in the secondary reactor. The residual sulphate data confirmed that the sulphide was not re-oxidised to sulphate and was partially oxidised toward elemental sulphur. There was a consistent increase in pH (Figure 9.19 B) observed within R2 which coincided with the converted sulphide and generation of alkalinity though BSR.

The relatively low sulphate reduction in the secondary reactor compared to the primary reactor was largely attributed to the degree of colonisation. During the start-up of the dual system the secondary reactor relied on the microbial activity of the overflow received from the primary reactor. Similarly observed during operation of the dual acetate-fed system, minimal sulphate reduction was achieved. It is possible that the SRB were inhibited by the exposure to the oxic zone during transport into the secondary reactor. The low sulphate reduction in R2 could be resolved through pre-colonisation with an active SRB culture, prior to dual operation. This would ensure that an active community is well developed and adapted for acetate utilisation toward sulphate reduction.

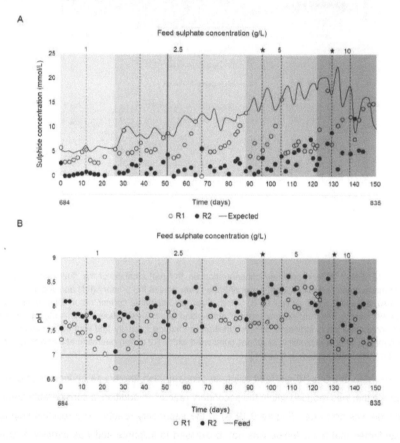

Figure 9.19: Effect of sulphate loading on the performance of the dual reactor showing the average A) dissolved sulphide concentration and B) pH of the effluent samples in the primary (R1) and secondary (R2) reactors, respectively. Biofilm disruption and harvest events are indicated by vertical dotted and solid lines, respectively. A change in feed sulphate concentration is indicated by the transition in shading intensity and was accompanied by a biofilm harvest. Additional biofilm disruptions conducted within R2 during operation at 5 and 10 g/L feed sulphate concentration are denoted by (★).

During the operation of the dual reactor system the average VFA concentration profiles obtained in the secondary reactor was compared to that observed in the primary reactor. At a

1 g/L feed sulphate concentration there was a decrease in the residual acetate concentration. Since complete lactate utilisation occurred within the primary reactor, the metabolism of acetate within the secondary reactor was favoured. However, there was minimal contribution toward sulphate reduction. A similar result was obtained in the dual acetate-fed system in Section 8.4, where relatively high acetate utilisation was observed in the secondary reactor with minimal decrease in sulphate concentration, which indicated that the metabolism of the acetate was performed by an active non-SRB population. The measured propionate concentration in the secondary reactor remained consistent with that observed in the primary reactor over the duration of the study. This suggests that minimal fermentation occurred within R2.

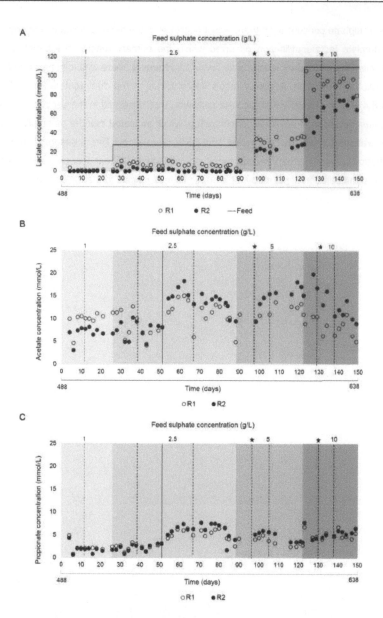

Figure 9.20: Effect of sulphate loading on the VFA profile in the dual reactor system showing A) lactate concentration B) acetate concentration and C) propionate concentration in the primary (R1) and secondary (R2) reactors. Biofilm disruption and harvest events are indicated by vertical dotted and solid lines, respectively. A change in feed sulphate concentration is indicated by the transition in shading intensity and was accompanied by a biofilm harvest. In addition, harvesting of the FSB occurred when sulphate concentration was changed. Additional biofilm disruptions conducted within R2 during operation at 5 and 10 g/L feed sulphate concentration is denoted by (★).

After increasing the feed sulphate concentration to 2.5 g/L there was increase in lactate concentration within the primary reactor. Lactate utilisation in the secondary reactor was low considering the high residual concentration that was still available. During operation of the dual system a large amount of colloidal sulphur and fragments of biofilm settled within the secondary reactor (Figure 9.21 B). The high sulphide generation and sulphur dispersion into the bulk volume favoured the formation of polysulphides which was evident by the characteristic yellow colour observed in Figure 9.21 B. A study by Mooruth (2013), highlighted the chemical reactions/processes that determine sulphur speciation and the dependence on pH and colloidal sulphur concentration in a sulphide oxidising dedicated LFCR. The study found that when the colloidal sulphur concentration was high (>2 mmol/L) and pH ranged between 8.1 and 9.5, polysulphides were produced. Similarly, within the current investigation, while operated at high feed sulphate concentration (>5 g/L) an excessive amount of colloidal sulphur was dispersed into the bulk volume. Despite the presence of the harvesting screen to collect the fragments of biofilm during disruption, a large amount of fine FSB particulates passed through the mesh and settled at the bottom of the reactor (Figure 9.21). The high sulphide concentration (15 mmol/L) in the bulk volume concomitant with an elevated pH of approximately 8.5 during operation at 5 g/L provided suitable conditions for the spontaneous formation of polysulphides.

Under alkaline conditions (pH > 8), elemental sulphur formed, either biologically or abiotically (Reaction 9.4), reacts with dissolved sulphide to form polysulphides of chain length (x) provided x ≥ 2 (Reaction 9.5). In addition, polysulphide can be generated through the reaction of dissolved sulphide with the free hydroxyl group (Reaction 9.6) generated during partial sulphide oxidation (Reaction 9.4) (van den Bosch, 2008).

$$HS^- + \tfrac{1}{2}O_2 \rightarrow S^0 + OH^- \qquad \text{(Reaction 9.4)}$$

$$HS^- + (x-1)S^0 \rightleftharpoons S_x^{2-} + H^+ \qquad \text{(Reaction 9.5)}$$

$$S_x^{2-} + xH_2O + (2x-2)e^- \rightleftharpoons xHS^- + xOH^- \qquad \text{(Reaction 9.6)}$$

The polysulphide in solution decreases dramatically with decreasing pH due to its instability at lower pH and therefore can be converted back its major constituents of elemental sulphur and sulphide through the addition of acidity (Findlay & Kamyshny, 2017; Boyd & Druschel, 2013). As a result, though not ideal, elemental sulphur incorporated within polysulphides can be recovered downstream through acidification. These results highlight the complexity of sulphur speciation within the hybrid LFCR which is highly dependent on the pH, colloidal sulphur concentration and hydroxyl ion concentration.

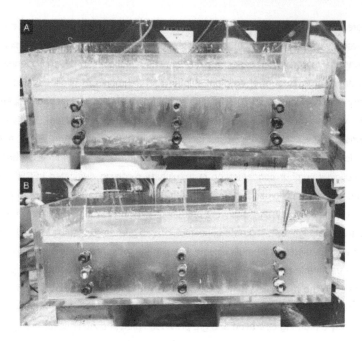

Figure 9.21: Photographs taken 24 hours after biofilm harvest showing the A) primary (R1) and B) secondary (R2) reactors. The distinctive yellow/green colouration associated with polysulphide formation as well as the presence of colloidal sulphur and settling fragments of biofilm within the secondary reactor can be observed.

9.3.2.3 Biofilm harvest and sulphur recovery in dual reactor system

A comparative analysis of the amount of biofilm harvested from the dual system from the respective reactors is shown in Figure 9.22. A similar increase in FSB recovery was observed in the secondary reactor as the feed sulphate concentration was increased from 2.5 to 5 g/L. Although the biofilm recovered from R2 was slightly higher compared to R1, during operation at 1 and 2.5 g/L, the higher proportion of FSB recovered during operation at 5 and 10 g/L was a consequence of the additional biofilm disruption events that was performed on days 96 and 129. This allowed the biofilm to regenerate outside of the biofilm disruption regime, applied to the primary reactor. The results demonstrated that an increase in the frequency of biofilm disruption can increase the overall sulphide conversion and sulphur recovery within the hybrid LFCR.

However, it should be noted that the high amount of sulphur generated as a result of the additional biofilm disruption event, contributed to the dispersion of colloidal sulphur into the bulk volume and the formation of polysulphides. Considering the loss of sulphur, it may be more beneficial to harvest the FSB rather than performing interim biofilm disruptions between harvesting. This would favour higher biofilm recovery by limiting the loss of sulphur and biofilm fragments to the bulk volume and effluent stream.

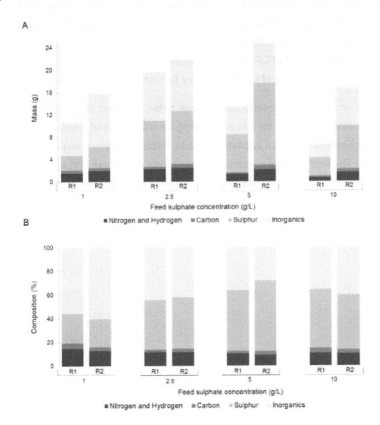

Figure 9.22: Effect of feed sulphate concentration on the floating sulphur biofilm recovery in the dual reactor system and the composition of nitrogen and hydrogen, carbon, sulphur and inorganics based on the A) amount (grams) of biofilm harvested and B) elemental composition (%) of the biofilm in the primary (R1) and secondary (R2) reactors of the dual hybrid LFCR system.

Elemental composition of the FSB recovered from R2 was consistent with the results obtained from R1 (Figure 9.22 B). A comparison of the elemental composition revealed that the FSB recovered from the respective reactors at each feed sulphate concentration, on average, exhibited a similar composition. However, there was an increasing trend in the proportion of elemental sulphur content (24 ± 0.5 to $56\pm5.8\%$) within the biofilm as the feed sulphate concentration increased from 1 to 5 g/L. Interestingly this coincided with the observed increase in VSRR (Table 9.3). In contrast, a decrease in the inorganic fraction (58±3.0 to 32±5.8%) was exhibited as the feed sulphate concentration increased.

The recovery of sulphur through harvesting of the FSB from the dual reactor system was evaluated and is summarised in Table 9.5. There was a marked increase in sulphur recovery, through the FSB, based on all three performance parameters (total sulphate-S load, expected sulphide-S and converted sulphide-S) when compared to single-stage operation (Table 9.5). The addition of the second LFCR unit increased the overall sulphide conversion (SO) and consequently the total amount of sulphur recovery through the harvested FSB.

Comparative analysis of the overall performance obtained during single and dual reactor operation revealed that, under the current operating conditions, a feed sulphate concentration of 2.5 g/L was optimal based on sulphate reduction, sulphide oxidation and sulphur recovery performance. The feed sulphate concentration is essential to promote high sulphate reduction activity and ensures that the sulphide to oxygen ratio favours the partial sulphide oxidation to elemental sulphur.

Table 9.5: Summary of the sulphur recovery in the 8 L dual reactor system as a function of feed sulphate concentration showing the amount of sulphur recovered through harvesting of the biofilm (FSB-S) as well as the respective sulphur recovery (%) based on each performance parameter including the total sulphate-S, expected sulphide-S (sulphate reduced: SR)) and converted sulphide-S (SO).

Sulphate concentration (g/L)	FSB-S harvested			Total S load [a]		Expected sulphide-S			Converted sulphide-S			
	R1 (g)	R2 (g)	Total (g)	Amount (g)	Recovery (%) [b]	SR (%)	Amount (g)	Recovery (%) [b]	SO (%)	Amount (g)	Recovery (%) [b]	Gap-S (g) [c]
					8 L lactate-fed (Dual system)							
1	2.6	3.7	6.3	19.1	33	42	8.0	79	85	6.8	93	0.5
2.5	8.2	9.5	17.7	67.8	26	49	33.2	53	79	26.1	68	8.4
5	6.8	14.6	21.4	124.7	17	36	44.5	48	68	30.1	71	8.7
10	3.3	7.7	11.0	212.7	5	10	21.3	52	61	13.0	85	0.2

[a] Total sulphur load over the duration of the experimental run
[b] Sulphur recovery based on total FSB-S
[c] Sulphur fraction released in the effluent

The recovery of sulphur through harvesting of the FSB increased substantially during operation of the dual reactor system. The ability to recovery elemental sulphur through harvesting of the FSB proved highly effective. During operation of the dual reactor system, sulphur recovery of the converted sulphide through harvesting of the FSB ranged between 68 and 93% (Table 9.5). In most sulphide removal processes the separation and recovery of elemental sulphur from the bioreactor contents is an essential stage to realise its value as a product for agricultural (fertiliser) and chemical (sulphuric acid) industry feedstock (Syed et al., 2006). Consequently, most sulphur recovery processes require an additional separation stage (Syed et al., 2006; Cai et al., 2017). The different separation techniques for elemental sulphur recovery was reviewed by Cai et al. (2017) and includes gravity sedimentation, membrane filtration, centrifugation and coagulation. Centrifugation is the most widely used technique and forms part of the Thiopaq™ process. The method is highly effective, capable of achieving >90% sulphur recovery, however, major disadvantage associated with the technique is the high cost (equipment and maintenance) and energy requirements (Cai et al., 2017). In this study the harvesting of the FSB proved highly effective and is advantageous over current separation methods.

9.3.2.4 Sulphur recovery through the effluent stream

The gap-S fraction which represents the converted sulphide not recovered through the FSB, was predominantly dispersed and suspended in solution as a colloidal fraction due to the hydrophilic nature of biologically produced sulphur (Cai et al., 2017). Furthermore, during operation there was a portion of biofilm fragments that detached from the FSB near the effluent port and was subsequently discharged within the effluent. Throughout this study the build-up of fine sulphur-like material in the effluent pipe over time often resulted in a blockage.

To validate that the gap-S fraction was largely elemental sulphur that was released into the effluent stream and pipe, the build-up of particulates within the effluent pipe was recovered for elemental analysis. A large amount of material (>3.5 g) with a similar consistency to that of the FSB was recovered from all three reactors (Figure 9.23). Upon further elemental analysis, it was revealed that it had similar composition to that of the FSB, containing a significant amount of elemental sulphur (approximately 30%). Described in Section 4.3, the conditions within the silicone effluent pipe provided oxygen-limiting conditions that favoured partial oxidation of the untreated sulphide to elemental sulphur. However, it is likely that the effluent pipe also accumulated a large fraction of fine colloidal sulphur particles and biofilm fragments originating from the effluent stream.

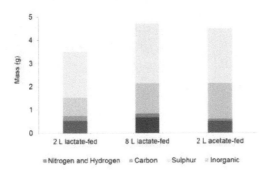

Figure 9.23: Recovery of elemental sulphur from the effluent pipe showing the total mass recovered and elemental composition of obtained from the 2 L and 8 L lactate-fed reactors as well as the 2 L acetate-fed reactor.

In addition to assessing the accumulation of sulphur in the effluent pipe by harvesting the build-up. The loss of sulphur to the effluent stream as colloidal sulphur and fragments of FSB, was further confirmed through HPLC analysis of the reactor and effluent samples (Figure 9.24). The sulphur concentration in the liquid sample was highly variable between samples, ranging between 0-6.5 mmol/L. As expected, the sulphur concentration was higher within the effluent compared to the reactor (bulk volume) samples. This was attributed to the fact that the effluent port is situated at the air-liquid interface where sulphide is continuously oxidised to elemental sulphur. As a result, fine fragments of the biofilm were often released into the overflow and dispersed into solution. These results confirm that the fraction of sulphur not recovered through the FSB (gap-S) was released into the effluent predominantly in the form of elemental sulphur during reactor operation. Similarly performed in conventional sulphide removal processes the gap-S fraction can be recovered through employing an additional separation method downstream to maximise sulphur recovery from the liquid phase (i.e. gravity sedimentation and inclined plate precipitation).

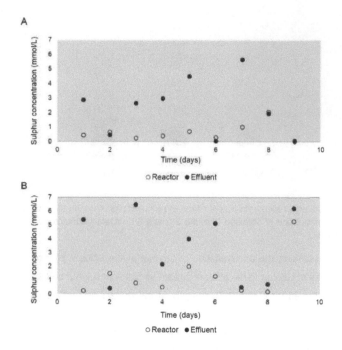

Figure 9.24: Sulphur concentration measured over time in the reactor volume and effluent of the A) 2 L and B) 8 L lactate-fed reactors.

9.4 Conclusion

This study investigated the impact of sulphate loading mediated through feed sulphate concentration on the performance of the hybrid LFCR. Based on comparative analysis across reactor geometry, both reactors were able to maintain relatively high sulphate reduction performance. The 2 L reactor demonstrated a higher resilience to operation at the high range feed sulphate concentration (10 g/L) compared to the 8 L reactor and was attributed to the degree of biomass which conferred higher system robustness, particularly on the SRB community. The increase in feed sulphate concentration resulted in an increase in sulphate reduction performance concomitant with an increase in sulphide oxidation and sulphur recovery through the FSB.

The increase in FSB recovery and sulphur composition indicated that the feed sulphate concentration is an essential parameter on the performance of the hybrid LFCR. The increased feed sulphate concentration resulted in higher sulphide availability for partial oxidation and subsequent increase in elemental sulphur deposit within the FSB. The increased

VSRR resulted in an increase in the rate at which the biofilm regenerated as well as the rate at which sulphide accumulated within the bulk volume due to the biofilm becoming oxygen limiting. To prevent sulphide inhibition on sulphate reduction the concentration of sulphide can be regulated by increasing the rate of biofilm disruption and harvesting.

Both reactors were able to maintain relatively high sulphate reduction over the duration of the study. However, there was a marked decrease in the sulphate conversion as the sulphate loading increased. The results reveal that the hybrid LFCR is capable of operation at high sulphate loading rates without considerable impact on process performance. Stoichiometric analysis concluded that incomplete oxidation of lactate was the predominant metabolic pathway however there was evidence of complete lactate oxidation which may have contributed to the overall BSR. This was attributed to the ability of the LFCR to retain biomass, facilitating the establishment of an active acetate-utilising community.

The operation of the dual reactor system proved effective with enhanced sulphide oxidation performance concomitant with an increase in the amount of sulphur recovery within the FSB. Furthermore, the operation of the secondary reactor demonstrated the potential for treatment of residual sulphate concentration and COD (lactate and acetate) received from the primary reactor.

Chapter 10

Conclusions & Recommendations

10.1 Introduction

The continued unrestrained discharge of sulphate-rich effluents originating from industrial activities, particularly in the mining sector, is a growing global concern and threat to freshwater security. The treatment of these effluents has been a subject of intensive research in recent years and the application of biological sulphate reduction mediated by SRB has been identified as a promising approach for its long term sustainable treatment. Its potential to effectively address all three main toxicological elements associated with ARD, namely sulphate, pH and heavy metals, with minimal sludge production makes it particularly attractive. Although many studies have evaluated sulphate reduction at laboratory-scale, the widespread application at large-scale implementation has been constrained by numerous challenges largely attributed to: 1) the management of the sulphide, 2) use of a cost effect electron donor, and 3) reaction kinetics of BSR.

In the current work, demonstration and characterisation of a novel semi-passive treatment technology capable of simultaneous biological sulphate reduction and partial sulphide oxidation with sulphur recovery was presented. The study involved the investigation into a range of operating conditions (reactor scale-up, electron donor, HRT, temperature, and sulphate loading) on process kinetics, stoichiometry, sulphur recovery and microbial community dynamics. These relationships can then be used to inform design of the process for optimal performance and process resilience. The driving factor behind the study was to develop a sustainable approach to treat ARD that overcomes the major challenges faced by BSR systems. The major findings of this work are presented in this final chapter.

10.2 Final Conclusions

The hybrid LFCR configuration was characterised by a fluid dynamic regime suitable for the integration of sulphate reduction and partial sulphide oxidation. The reactor was primarily governed by advective and diffusive mixing with minimal turbulence at the surface, which facilitated the partitioning of a discrete aerobic zone at the surface and anaerobic zones within the bulk volume of the reactor. The results achieved in Chapter 4 supported the Hypothesis 1 which successfully demonstrated the simultaneous sulphate reduction and sulphide oxidation through the formation of a structurally sound FSB with sulphur recovery within a single reactor unit. The carbon microfibres were an important feature of the hybrid LFCR and ensured effective attachment of SRB to achieve high sulphate reduction rates. The addition of the harvesting screen proved highly effective for recovering the FSB with minimal disturbance to the system. The study achieved high sulphate reduction and sulphide removal concomitant with the recovery of elemental sulphur-rich biofilm. A key operating feature of the hybrid LFCR was the management of the FSB. The intermittent disruption and harvesting regime had an impact on the sulphate reduction and sulphide oxidation process. The study found that the regulation of the biofilm disruption regime is essential to achieve optimal sulphur recovery,.

Stoichiometric analysis of the lactate-fed reactor revealed that incomplete lactate oxidation was favoured although complete lactate oxidation was present, particularly during operation at longer HRT. This was attributed to the ability of the LFCRs to retain and develop an active acetate utilising community. These results were distinctly different from those reported in literature where lactate is often limited by incomplete oxidation to acetate with minimal complete oxidation (Celis et al., 2013). In the current study the operation at a 0.7 COD/sulphate ratio and low dilution rate as well as the ability of the LFCR to retain high biomass favoured the activity and proliferation of complete oxidisers that included both SRB and non-SRB.

Simulation of the pilot plant dimensions in a laboratory scale 8 L LFCR was evaluated in order to determine effects of change in reactor geometry and volume on fluid dynamics and process performance. The reactor was operated alongside the 2 L LFCR. The results demonstrated the robustness of the hybrid LFCR process when scaled-up by a factor of 4. The comparative assessment of the 2 L and 8 L reactors showed minimal difference in overall process performance and were relatively consistent across operating conditions evaluated.

The provision of a suitable cost effective electron donor has been identified as one of the biggest challenges facing the industrial application of biological sulphate reducing technologies. In the current work the use of lactate which served as an effective electron donor achieved high sulphate reduction for evaluating the hybrid LFCR process. Although it proved effective in the operation of the hybrid LFCR, the use of lactate is not a preferred carbon source when considering industrial application due to availability and cost, as well as its tendency not to achieve complete oxidation of all the carbon source, leaving residual COD as acetate in the effluent. Acetate was identified as an alternative carbon source. Throughout the study, the acetate-fed system showed potential as a source carbon and was comparable to the lactate-fed reactors. The system was, however, more sensitive to change in operating conditions, including oxygen exposure after biofilm disruption, low temperature and high dilution rates. This resulted in lower process resilience than the lactate-operated system. As a result, the system required longer periods of operation between biofilm disruption to recover sulphate reduction performance which limited the frequency of FSB harvest and sulphur recovery. The higher sensitivity to operational perturbation was largely attributed to the microbial community. Complete oxidisers (acetate utilising SRB) are characterised by a slow growth rate and are physiologically and metabolically less versatile than incomplete oxidisers. Despite this constraint, the system demonstrated increased robustness and resilience over time as biomass accumulated within the reactors.

The results observed during the exposure to change in operating conditions based on sulphate loading, mediated by HRT and feed sulphate concentration, as well as temperature were promising. For efficient operation of the hybrid LFCR, a 2 day HRT was identified as optimal achieving 0.144 and 0.129 mmol/L.h in the lactate-fed reactors and 0.115 mmol/L.h in the acetate-fed reactor. This was equivalent to a sulphate conversion of 66, 60, and 53%, respectively. For all three reactors supplemented with lactate or acetate, although sulphate reduction gradually decreased as temperature decreased, the critical temperature (T_{crit}) was identified at 15°C, below which (10°C) sulphate reduction decreased. This was largely attributed to the cold induced stress that was experienced, which affected overall metabolic activity. At 10°C the 2 L and 8 L lactate-fed reactors were able to maintain a VSRR of 0.059 and 0.042 mmol/L.h with a corresponding conversion of 27 and 20%, respectively. The acetate-fed reactor was highly sensitive to low temperature at 10°C, achieving a VSRR of only 0.010 mmol/L.h with a corresponding conversion of 5%. An increase in sulphate loading during operation at 2.5 g/L ensured efficient performance of the sulphide oxidation component with increased sulphur recovery in the FSB, across both lactate-fed reactors. The highest VSRR recorded during the effect of feed sulphate concentration study was 0.159 mmol/L.h with a conversion of 59% in the 2 L lactate-fed reactor during operation at a feed sulphate concentration of 2.5 g/L, while a 0.200 mmol/L.h VSRR was achieving in the 8 L lactate-fed reactor with a conversion of 34% at a 5 g/L feed sulphate concentration. The data obtained from these studies supported the Hypothesis 2, which focused on further characterisation of the hybrid process as a function of operating conditions.

From a practical perspective the hybrid LFCR is to be operated as a semi-passive process without the control of temperature and feed sulphate concentration at industrial scale. Therefore, the regulation of the operating HRT becomes the most critical management parameter to ensure satisfactory treatment. Under stress conditions of temperature and feed sulphate concentration, kinetic constraints can be overcome by operating at a longer HRT. The effect of HRT on the sulphate reduction revealed that system performance is governed by a compromise between VSRR and sulphate conversion. As HRT is decreased and VSRR increases with increasing VSLR, the sulphate conversion decreased. Therefore, depending on the application and the desired quality of water, the choice of HRT should consider both performance values to facilitate optimal process performance.

The hybrid LFCR hosts a rich source of complex microbial communities within discrete environments (attached biofilm, planktonic phase, FSB and associated phase at the interfacial zone) that facilitated the development of active SRB and SOB communities within a single unit of operation. Investigation into the microbial community dynamics revealed the preferential attachment of SRB and fermentative microorganisms onto the carbon microfibers. Key SRB genera detected in this study were dominantly represented by *Desulfovibrio* and *Desulfomicrobium* within the lactate-fed reactor and *Desulfobacter* within the acetate-fed reactor. Other notable SRB that were detected at lower abundance were *Desulfocurvus*, *Desulfarculus*, *Ruminoccoccacae*. During the study of HRT on reactor performance, the microbial communities shifted when dilution rate was decreased from 5 to 2 days. The increase in abundance of fermentative genera (*Veilonella*, *Synergistetes* and *Bacteroidetes*) was concomitant with the observed increase in lactate fermentation. In contrast the high abundance of SRB genera (*Desulfovibrio* and *Desulfomicrobium*) at low dilution rates coincided with high sulphate reduction performance. The FSB represented a unique ecological environment, harbouring a taxonomically diverse SOB community that requires further exploration. The major SOB taxonomic phyla detected in the current study included Alphaproteobacteria, Betaproteobacteria, Gammaproteobacteria, Epsilonproteobacteria and Chlorobi. While the microbial communities were largely similar within the lactate-fed reactors, there were distinct differences observed within the acetate-fed reactor this included the releative abundance of major phyla such as Bacteroidetes, Synergistetes and Deltaproteobacteria. The outcome of this study revealed that the microbial community composition was more strongly influenced using different carbon substrate than by reactor geometry and scale up of the hybrid LFCR. The LFCR represents a unique environment in which the sulphur cycle has been effectively established to achieve water treatment and harvest of products. Few technologies and natural environments exist in which communities of both SRB and SOB occur simultaneously within close proximity. This is of great significance not only from an engineering perspective for process design and implementation, but also toward microbial community studies of the sulphur cycle. This work successfully validated Hypothesis 3, which showed that the use of different electron donor selects for distinctly different microbial communities that are essential for process performance. Although the microbial community was also evaluated as a function of HRT, reactor geometry, the use of a different electron donor proved most influential on the microbial community composition within the reactors.

The operation of a dual reactor system proved effective for enhancement of sulphide oxidation and sulphur recovery. The addition of a second LFCR unit in series increased the amount of FSB recovered and consequently the sulphur recovery efficiency. Furthermore, the secondary reactor facilitated the removal of residual acetate and lactate. During dual operation of the 2 L acetate-fed dual reactor system, the system was exposed to low temperature which largely limited the development and colonisation of the second reactor in both operating zones for enhanced sulphate reduction. In the 8 L lactate-fed dual reactor system, sulphate reduction within the secondary reactor gradually increased as biomass accumulated and SRB established, while the rapid formation of the FSB was consistent with that observed in the primary reactor.

Over the duration of this work the reactors were operated continuously over an extended period and were exposed to a range of operating conditions. Upon completion of this work the 2 L lactate-fed reactor had been operated for a total of 1484 days while the 8 L reactor and 2 L acetate-fed reactor was operated for 835 and 732 days, respectively. The extent of continuous operation and assessment of multiple operating conditions, shown in this work, had not been report elsewhere. Most BSR studies are operated at laboratory-scale for short periods of time during which a single operating condition is evaluated. These studies are often inoculated with a highly active SRB culture and operated for a relatively short period without allowing the reactors to operate long enough for the microbial community to fully colonise and establish. Results from this work demonstrated the long-term viability and robustness of the hybrid LFCR process as well as its resilience to recover stable performance.

It is envisaged that this present research will contribute to the development and successful application of a novel wastewater treatment to address the persistent low flow ARD associated with diffuse sources. Furthermore, a combination of the process kinetics, reaction stoichiometry and the microbial community dynamics reported in this study has potential for use as a diagnostic tool in the assessment of the hybrid LFCR process. This work also provides insight into strategies to maximise the efficiency of process performance. Although the technology offers a promising approach to addressing low volume sulphate-rich mine impacted water, it is not limited to only ARD and could be applied for treating low volume industrial effluents containing high sulphate concentrations. The recovery of sulphur as a value product is an attractive feature of the process, requiring minimal energy input or maintenance and lends to the feasibility of the process. The sulphur is highly applicable for agricultural amelioration and chemical production industry while the treated effluent can be applied for fit for purpose/ irrigation. The process represents a potential strategy toward achieving sustainable mining and developing a circular economy approach to addressing mine impacted water (Figure 10.1).

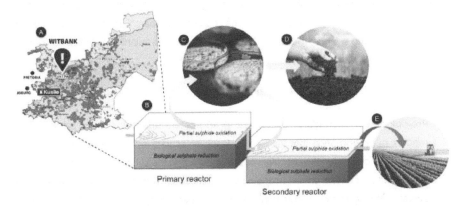

Figure 10.1: Potential application of the dual hybrid LFCR process treating low volume ARD originating from A) coal diffuse sources (Baillie, n.d.), B) The dual reactor system is capable of effectively removing sulphate and management of the generated sulphide through the C) recovery of elemental sulphur. The sulphur can be applied in D) agriculture as a fertiliser while the E) treated effluent can be used for irrigation.

Key advantages demonstrated by the hybrid LFCR process in this work include:

- Semi-passive operation (minimal energy input, maintenance, control of oxygen conditions)
- Simultaneous sulphate reduction and sulphide removal with elemental sulphur recovery
- Effective FSB harvesting system for sulphur separation and recovery
- High biomass retention and attachment of carbon microfibers
- Complete oxidation of lactate toward sulphate reduction
- Increased pH through bicarbonate (SR) and hydroxyl ion (SO) production

- Stability of system to change in operational conditions (HRT, temperature and sulphate loading)
- Robustness to scale-up by a factor of 4
- Long term viability and system resilience to recovery performance

10.3 Recommendations for future work

Although in this study the hybrid LFCR system was extensively evaluated as a function of operating conditions, there is still a need for further research and development towards achieving optimal performance. Based on the outcomes of this study, the following areas of work are recommended for further investigation to generate additional data that can be used to improve the hybrid LFCR process. The potential research is categorised into 1) Enhancement of the sulphate reduction, 2) Enhancement of the sulphide oxidation component and 3) Developmental and application of the hybrid LFCR.

10.3.1 Enhancement of biological sulphate reduction component

1. **Pre-colonise a secondary reactor using an acetate adapted culture prior to dual reactor operation** such that the accumulation of acetate within the primary reactor is effectively metabolised toward sulphate reduction within the secondary reactor, increasing the efficiency of lactate utilisation and sulphate reduction. The pre-colonisation of an active acetate utilising microbial community within the secondary reactor could enhance the utilisation of residual acetate toward sulphate reduction.

2. **Investigate different configuration or orientation of carbon microfibers** as well as the use of alternative support material for biomass attachment within the reactor with better distribution throughout the bulk volume. Ideally the support matrix should increase the capacity of the LFCR to retain biomass through providing enhanced surface area for biomass attachment with minimal impact on the working volume. In addition, the support material should be effective in biomass distribution throughout the bulk volume to increase contact and exposure of the organic substrate with the active population.

3. **Evaluate the use of an alternative carbon source.** Propionate which is the second major component released during anaerobic digestion, has been shown to be highly preferential for SRB growth and activity. Since the metabolism of propionate is largely limited to a select group of microorganisms, the competition for the substrate is less problematic and

may serve as an effective source of carbon. Although, the current study demonstrated the technical feasibility of the hybrid LFCR, the challenges faced by the provision of a readily available, cost effective electron donor still needs to be addressed. Studies should explore the use of cheap organic-rich waste streams that could potentially be used for sulphate reduction.

10.3.2 Enhancement of partial sulphide oxidation component:

1. **Maximise the sulphur recovery through optimally regulating the biofilm disruption regime.** The performance of the sulphide oxidation was largely limited by the biofilm disruption and harvesting. The performance can be increased, particularly within the secondary reactor by regulating more frequent harvesting of the biofilm. Since the primary reactor is essential for effective sulphate reduction, this process should not be compromised by frequent FSB harvesting.

10.3.3 Development and application of the hybrid LFCR

1. **Evaluate the effect of operating conditions such as pH, heavy metals and use of raw ARD feed and other sulphate or sulphide-rich effluents (i.e. tannery effluent).** For further application and improvement of the LFCR process, it is crucial that the system is adapted and exposed to a more complex and fluctuating influent composition such as raw ARD discharge or industrial effluents characterised by high sulphate concentration. Additionally, it is recommended that the effects of acidic pH range and heavy metal concentration on process performance be evaluated. Low pH and the presence of heavy metals is often associated with ARD streams and can be inhibitory to BSR. Hence evaluating its effect on the hybrid LFCR is important toward the feasibility of the process and to what extent a neutralisation and metal removal step upstream of the hybrid LFCR is required.

2. **Investigation into microbial activity at a functional level.** Microbial community dynamics played a critical role on the overall performance of the system. Further investigation into the microbial ecology when exposed to changes in operating conditions such as temperature and sulphate loading are essential to understanding the relationship between process performance and community dynamics. Although 16S rRNA methods can be used to determine composition and relative abundance of different populations within the reactors, the method is limited to taxonomic classification and does not reflect

the active microbial population. With the latest advancement in next generation sequencing and whole genome sequencing, understanding the metabolic potential of the microbial communities can be elucidated. These studies will provide a complete overview of the metabolic reactions occurring within the system and may reveal new insight into the cycling of carbon and sulphur within the hybrid LFCR. The occurrence of the FSB and its functionality with regards to biofilm organisation, localisation and function of the microbial community that make up its structure is still under developed and requires further research; whole genome sequencing is desirable to provide insight into the functional capacities as well as community structure.

References

Abba, S.I., & Elkiran, G. 2017. Effluent prediction of chemical oxygen demand from the wastewater treatment plant using artificial neural network application, *Procedia Computer Science*, 120:156-163.

Al-zuhair, S., El-naas, M. H., & Al-hassani, H. 2008. Sulphate inhibition effect on sulphate reducing bacteria. *Journal of Biochemestry Technology*. 1:39–44.

Amend, J.P., & Teske, A. 2005. Expanding frontiers in deep subsurface microbiology. *Palaeogeography, Palaeoclimatology, Palaeoecology*. 219:131-155.

Aoyagi, T., Hamai, T., Hori, T., Sato, Y., Kobayashi, M., Sato, Y., Inaba, T., Ogata, A., Habe, H., & Sakata, T. 2018. Microbial community analysis of sulphate reducing passive bioreactor for treating acid mine drainage under failure conditions after long-term continuous operation. *Journal of Environmental and Chemical. Engineering*. 6:5795–5800.

APHA, 2012. *Standard methods for the examination of water and wastewater*, 22nd edition. E. W. Rice, R. B. Baird, A. D. Eaton and L. S. Clesceri, Eds. American Public Health Association (APHA), American Water Works Association (AWWA) and Water Environment Federation (WEF). USA: Washington, D.C.

Armitano, J., Mejean, V., & Jourlin-Castelli, C. 2013. Aerotaxis governs floating biofilm formation in *Shewanella oneidensis*. *Environmental Microbiology*. 11:3108-18.

Arnold, M., Gericke, M., & Muhlbauer, R. 2016. Technologies for sulphate removal with valorisation options. *Proceedings of the IMWA symposium: Mining Meets Water – Conflicts and Solutions*. 11- 15 July 2016. Leipzig, Germany. 1343-1346.

Ashe, N.L., McLean, I., & Nodwell, M. 2008. Review of operations of the biosulphide process plant at the copper Queen mine, Bisbee, Arizona. *Proceedings of the 6th international symposium on hydrometallurgy, Society for Mining, Metallurgy and Exploration*. 2008. Phoenix: Arizona, 98–107.

Ayangbenro A.S., Olanrewaju O.S., & Babalola O.O. 2018. Sulphate-Reducing Bacteria as an Effective Tool for Sustainable Acid Mine Bioremediation. *Frontiers in Microbiology.* 9:1986.

Bade, K., Manz, W., & Szewzyk, U. 2000. Behaviour of sulphate reducing bacteria under oligotrophic conditions and oxygen stress in particle-free systems related to drinking water, *FEMS Microbiology Ecology.* 32(3):215–223.

Bai, H., Kang, Y., Quan, H., Han, Y., Sun, J., & Feng, Y. 2013. Treatment of acid mine drainage by sulphate reducing bacteria with iron in bench scale runs. *Bioresource Technology.* 128(0):818-822.

Bai, H., Liao, S., Wang, A., Huang, J., Shu, W., & Ye, J. 2019. High-efficiency inorganic nitrogen removal by newly isolated *Pannonibacter phragmitetus* B1. *Bioresource Technology.* 271:91-99.

Baillie, M. n.d. *Poisoned people via Greenpeace Africa.* Available: http://www.green-humanity.com/poisoned-people.html [2019, July 25].

Banks, D., Younger, P.L., Arnesen, R.T., Iversen, E.R., & Banks, S.B. 1997. Mine-water chemistry: the good, the bad and the ugly. *Environmental Geology.* 32(3):157-174.

Baronofsky, J.J., Schreurs, W.J.A., & Kashket, E.R. 1984. Uncoupling by acetic acid limits growth of and acetogenesis by *Clostridium thermoaceticum*. *Applied Environmental Microbiology.* 48:1134-1139.

Barton L.L., Fardeau ML., & Fauque G.D. (2014) Hydrogen Sulfide: A Toxic Gas Produced by Dissimilatory Sulfate and Sulfur Reduction and Consumed by Microbial Oxidation. In: Kroneck P., Torres M. (eds) The Metal-Driven Biogeochemistry of Gaseous Compounds in the Environment. *Metal Ions in Life Sciences.* 14. Springer, Dordrecht.Barton, L.L., & Fauque, G.D. 2009. Biochemistry, Physiology and Biotechnology of Sulphate Reducing Bacteria. *Advances in Applied Microbiology.* 68:41–98.

Baskaran, V., & Nemati, M. 2006. Anaerobic reduction of sulphate in immobilised cell bioreactors, using a microbial culture originated from an oil reservoir. *Biochemical Engineering Journal*, 31(2):148–159.

Berben, T., Overmars, L., Sorokin, D. Y., Muyzer, G., & Bailey, J. 2017. Comparative genome analysis of three thiocyanate oxidising *Thioalkalivibrio* species isolated from soda lakes. *Frontiers in Microbiology.* 8:1–14.

Bertolino, S.M., Rodrigues, I.C.B., Guerra-Sá, R., Aquino, S.F., & Leão, V. 2012. Implications of volatile fatty acid profile on the metabolic pathway during continuous sulphate reduction. *Journal of Environmental Management*. 103:15–23.

Bezuidenhout, N., Verburg, R., Chatwin, T., Ferguson, K. 2009. INAP's global acid rock drainage guide and the current state of Acid Rock Drainage assessment and management in South Africa. *International Mine Water Conference Proceedings*, 19 – 13th October 2009, Pretoria, South Africa.

Bisht, S.C. 2011. *Effect of low temperature on bacterial growth*. Available: https://www.biotecharticles.com/Biology-Article/Effect-of-Low-Temperature-on-Bacterial-Growth-721.html [2019, July 23].

Botes, L., Price, B., Waldron, M., & Pitcher, G.C. 2002. A simple and rapid scanning electron microscope preparative technique for delicate "gymnodinioid" dinoflagellates. *Microscopy Research and Technique*. 59:128-130.

Bowell, R.J. 2004. A review of sulphate removal options for mine waters. – In: Jarvis, A. P., Dudgeon, B. A. & Younger, P. L. (eds): *Proceedings International Mine Water Association Symposium*. 75-91, Newcastle, United Kingdom.

Boyd, E.S., & Druschel, G.K. 2013. Involvement of intermediate sulfur species in biological reduction of elemental sulfur under acidic hydrothermal conditions. *Applied and Environmental Microbiology*. 79(6):2061-2068.

Brahmacharimayum, B., Mohanty, M.P., & Ghosh, P.K. 2019. Theoretical and practical aspects of biological sulphate reduction: a review. *Global NEST Journal*. 21(2):222-224.

Bruser, T., Lens, P.N.L., & Truper, H.G. 2000. The biological sulphur cycle. In *Environmental technologies to treat sulfur pollution: Principles and Engineering*. P.N.L. Lens & P. Holsof Pol, Eds. London: IWA Publishing. 47-86.

Buisman C.J.N., Huisman, J., Dijkman, H., & Bijmans, M.F.M. 2007. Trends in application of industrial sulphate reduction for sulfur and metal recycling. *Proceedings of the European Metallurgical Conference on the Horizons of Sustainable Growth of the Non-Ferrous Metals Production*. 2007. Dusseldorf, Germany. 383–387.

Buisman, C., Post, R., Ijspeert, P., Geraats, G., & Lettinga, G. 2010. Biotechnological process for sulphide removal with sulphur reclamation. *Engineering in Life Sciences*. 9:255–267.

Buisman, C.J., Geraats, B.G., Ijspeert, P., & Lettinga, G. 1990. Optimization of sulphur production in a biotechnological sulphide-removing reactor. *Biotechnology and Bioengineering.* 35: 50–56

Cai, J., Zheng, P., Qaisar, M., & Zhang, J., 2017. Elemental sulfur recovery of biological sulfide removal process from wastewater: A review. *Critical Reviews in Environmental Science and Technology.* 3389:1–21.

Camiloti, P.R., Oliveira, G.H.D., & Zaiat, M. 2016. Sulfur recovery from wastewater using a micro-aerobic external silicone membrane reactor (ESMR). *Water, Air, & Soil Pollution.* 227:31.

Caporaso, J.G., Kuczynski, J., & Stombaugh, J. 2010. QIIME allows analysis of high throughput community sequencing data. *Nature Methods.* 7(5): 335–336.

Carbonero, F., Benefiel, A.C., Alizadeh-Ghamsari A.H., & Gaskins, H.R. 2012. Microbial pathways in colonic sulfur metabolism and links with health and disease. *Frontiers in Physiology.* 3(448):1-11.

Carlson, H.K., Iain, C.C., Blazewicz, S.J., Iavarone, A.T., & Coates, J.D. 2013. Fe(II) oxidation is an innate capability of nitrate reducing bacteria that involves abiotic and biotic reactions. *Journal of Bacteriology.* 195(14):3260-3268.

Carrondo, M.J.T., Silva, J.M.C., Figueira, M.I.I., Ganho, R.M.B., & Oliveira, J.F.S. 1983. Anaerobic filter treatment of molasses fermentation wastewater. *Water Science Technology.* 15:117-28.

Celis, L.B., Villa-Gómez, D., Alpuche-Solís, A.G., Ortega-Morales, B.O., & Razo-Flores, E. 2009. Characterisation of sulphate-reducing bacteria dominated surface communities during start-up of a down-flow fluidised bed reactor. *Journal of Industrial Microbiology and Biotechnology.* 36:111–121.

Celis, L. B., Gallegos-Garcia, M., Vidriales, G., & Razo-Flores, E. 2013. Rapid start-up of a sulfidogenic biofilm reactor: Overcoming low acetate consumption. *Journal of Chemical Technology and Biotechnology.* 88(9): 1672–1679.

Chabane, Y.N., Marti, S., Rihouey, C., Alexandre, S., Hardouin, J., et al. 2014. Characterisation of Pellicles Formed by Acinetobacter baumannii at the Air-Liquid Interface. *PLoS ONE.* 9(10):e111660.

Chalima, A., Oliver, L., de Castro, L.F., Karnaouri, T.D., & Topakas, E. 2017. Utilisation of volatile fatty acids from microalgae for the production of high added value compounds. *Fermentation*. 3(4):54.

Chen., S., & Dong, X. 2005. *Proteiniphilum acetatigenes* gen. nov., sp. nov., from a UASB reactor treating brewery wastewater. *International Journal of Systematic Evolutionary Microbiology*. 55(6):2257-61.

Chuichulcherm, S., Nagpal, S., Peeva, L., & Livingston, A. 2001. Treatment of metal-containing wastewaters with a novel extractive membrane reactor using sulphate reducing bacteria. *Journal of Chemical Technology and Biotechnology*. 76:61–68.

Cline, C., Hoksberg, A., Abry, R., & Janssen, A. 2003. Biological process for H_2S removal from gas streams the shell-paques / thiopaq™ gas desulfurisation process. *Proceedings of the 53rd Laurance Reid Gas Conditioning Conference*. 23-26 February 2003. Oklahoma, USA. 23–26.

Coetser, S., Heath, R., Molwantwa, J., Rose, P., & Pulles, W. 2005. Implementing the degrading packed bed reactor technology and verifying the long term performance of passive treatment plants at Vryheid Coronation Colliery. (WRC Report; no. 1348/1/05). South Africa: Water Research Commission.

Colleran, E., Finnegan, S., & Lens, P. 1995. Anaerobic treatment of sulphate-containing waste streams. *Antonie Van Leeuwenhoek*. 67(1):29-46.

Cui, J., Chen, X.P., Nie, M., Fang, S.B., Tang, B.P., Quan, Z.X., Li, B., & Fang, C.M. 2017. Effects of *Spartina alterniflora* invasion on the abundance, diversity, and community structure of sulphate reducing bacteria along a successional gradient of coastal salt marshes in China. *Wetlands*. 37(2):221–232.

Cypionka, H. 2000. Oxygen respiration by Desulfoviyxbrio species. *Annual Review of Microbiology*. 54: 827-848.

Dar, S.A., Yao, L., Van Dongen, U., Kuenen, J.G., & Muyzer, G. 2007. Analysis of diversity and activity of sulphate-reducing bacterial communities in sulfidogenic bioreactors using 16S rRNA and *dsrB* genes as molecular markers. *Applied Environmental Microbiology*. 73:594–604.

Dessi, P., Porca, E., Haavisto, J., & Lakaniemi, A. 2018. Composition and role of the attached and planktonic microbial communities in mesophilic and thermophilic xylose-fed microbial fuel cells *RSC Advances*. 8(6):3069–3080.

Dijkman, H. 1995. Biological gas desulphurisation. *Mededelingen- faculteit landbouwkundige en toegepaste biologische wetenschappen.* 60(4):2677-2684.

Dogan, E. C., Turker, M., Dagasan, L., & Arslan, A. 2012. Simultaneous sulfide and nitrite removal from industrial wastewaters under denitrifying conditions. *Biotechnology and Bioprocess Engineering* 17:661–668.

Dolla, A., Fournier, M., & Dermoun, Z. 2006. Oxygen defense in sulphate- reducing bacteria. *Journal of Biotechnology.* 126(1):87–100.

Drury, W.J. 2000. Modelling of sulphate reduction in anaerobic solid substrate bioreactors for mine drainage treatment. *Mine Water and the Environment.* 19(1):19–29.

Dubourg, G., Lagier, J.C., Hue, S., Surenaud, M., Bachar, D., et al. 2016. Gut microbiota associated with HIV infection is significantly enriched in bacteria tolerant to oxygen. *BMJ Open Gastroenterol.* 3(1):1–9.

Dvorak, D.H., Hedin, R.S., Edenborn, H.M., & McIntire, P.E. 2004. Treatment of metal contaminated water using bacterial sulphate reduction: Results from pilot-scale reactors. *Biotechnology and Bioengineering.* 40(5):609 -616.

Edgar, R.C. 2010. Search and clustering orders of magnitude faster than BLAST. *Bioinformatics.* 26:2460–1.

Elkanzi, E.M., 2009. Simulation of the Process of Biological Removal of Hydrogen Sulfide from Gas, Proceedings of the 1st Annual Gas Processing Symposium. *Advances in Gas Processing.* 1, 266-275.

Elliott, P., Ragusa, S., & Catcheside, D. 1998. Growth of sulphate-reducing bacteria under acidic conditions in an upflow anaerobic bioreactor as a treatment system for acid mine drainage. *Water Research.* 32(12):3724–3730.

Elsner, M.P., Menge, M., Müller, C., & Agar, D.W. 2003. The Claus process: teaching an old dog new tricks. *Catalysis Today.* 79:487-494.

El-Tarabily, K., Soaud, A, Saleh, M., & Matsumoto, S. 2006. Isolation and characterisation of sulfur-oxidising bacteria, including strains of Rhizobium, from calcareous sandy soils and their effects on nutrient uptake and growth of maize (*Zea mays L.*). *Australian Journal of Agricultural Research.* 57(1):101-111.

Erasmus, C.L. 2000. A preliminary investigation of the kinetics of biological sulphate reduction using ethanol as a carbon source and electron donor. MSc(Eng) dissertation. University of Cape Town, South Africa.

European Association for Specialty Yeast Products. 2015. Yeast Extract. European Association for Speciality Yeast Products, 1–2.

Feris, L., & Kotze, L.J. 2014. The regulation of acid mine drainage in South Africa: Law and governance perspectives. *Potchefstroomse Elekroniese Regsblad*. 17(5): 2105-2163.

Ferrentino, R., Langone, M., & Andreottola, G. 2017. Temperature Effects on the activity of denitrifying phosphate accumulating microorganisms and sulphate reducing bacteria in anaerobic side-stream reactor. *Journal of Environment and Bio Research*. 1(1):1-7.

Findlay, A., & Kamyshny, A. 2017. Turnover rates of intermediate sulfur species (S_x^{2-}, S^0, $S_2O_3^{2-}$, $S_4O_6^{2-}$, SO_3^{2-}) in anoxic freshwater and sediments. *Frontiers in Microbiology*. 8:2551.

Galiana-Aleixandre, M.V., Mendoza-Roca, J.A., & Bes-Pia, A. 2011. Reducing sulphates concentration in the tannery effluent by applying pollution prevention techniques and nanofiltration. *Journal of Cleaner Production*. 19(1):91-98.

Gazea, B., Adam, K., & Kontopoulous, A. 1996. A review of passive systems for the treatment of acid mine drainage. *Minerals Engineering*. 9:23-42.

Ghosh, W., & Dam, B. 2009. Biochemistry and molecular biology of lithotrophic sulfur oxidation by taxonomically and ecologically diverse bacteria and archaea. *FEMS Microbiology Reviews*. 33(6):999-1043.

Glombitza, F. 2001. Treatment of acid lignite mine flooding water by means of microbial sulphate reduction. *Waste Management*. 21:197-203.

Gomes, L., & Mergulhão, F. 2017. SEM Analysis of Surface Impact on Biofilm Antibiotic Treatment. *Scanning*. 1-7.

Gopi Kiran, M., Pakshirajan, K., & Das, G., 2017. An overview of sulfidogenic biological reactors for the simultaneous treatment of sulphate and heavy metal rich wastewater. *Chemical Engineering Science*. 58: 606-620.

Greben, H. A., & Maree, J. P. 2000. The effect of reactor type and residence time on biological sulphate and sulphide removal rates. *Proceedings of the WISA 2000 Biennial Conference*. 28 May - 1 June 2000. Sun City, South Africa.

Greben, H.A., Bologo, H., & Maree, J.P. 2002. The effect of different parameters on the biological volumetric and specific sulphate removal rates. Water SA. Special Edition, *WISA Biennial Conference*. 33.

Greben, H.A., Tjatji, M., & Maree, J.P. (2004) Biological sulphate reduction at different feed COD/SO4 ratios using propionate and acetate as the energy source. *Proceeding of the 9th International Mine Water Association Symposium*. Newcastle, United Kingdom, 101-109.

Guerrero, L., Montalvo, S., Huilinir, C., Campos, J.L., Barahona, A., & Borja, R., 2016. Advances in the biological removal of sulphides from aqueous phase in anaerobic processes: a review. *Environment Review*. 24: 84–100.

Günther, P., & Mey, W. 2008. Selection of mine water treatment technologies for the Emalahleni (Witbank) Water Reclamation Project. *Proceedings of the WISA Biennial Conference*. 18-22nd May 2008. Sun City, South Africa. 1-14.

Gupta, A., Dutta, A., Sarkar, J., Panigrahi, M.K., & Sar, P. 2018. Low-Abundance Members of the Firmicutes Facilitate Bioremediation of Soil Impacted by Highly Acidic Mine Drainage From the. *Frontiers in Microbiology*. 9:1–18.

Habets, L.H.A., & de Vegt, A.L. 1991. Anaerobic treatment of bleached TMP and CTMP effluent in the biopaq UASB system. *Water Science & Technology*. 24(3-4):331-345.

Hao, O.J., Chen, J.M., Huang, L., & Buglass, R.L. 1996. Sulphate-reducing bacteria. Critical Reviews in *Environmental Science and Technology*. 26:155-187.

Hao, T., Xiang, P., Mackey, H.R., Chi, K., Lu, H., Chui, H., van Loosdrecht, M.C.M., & Chen, G,H. 2014. A review of biological sulphate conversions in wastewater treatment. *Water Research*. 65:1–21.

Harrison, S.T.L., van Hille, R.P., Mokone, T., Motleleng, L., Smart, M., Legrand, C., Marais, T. 2014. Addressing the Challenges Facing Biological Sulphate Reduction as a Strategy for AMD Treatment: Analysis of the reactor stage: raw materials products and process kinetics. (WRC report; no. 2110/1/14). South Africa: Water Research Commission. ISBN: 978-1-4312-0605-6.

Henshaw, P., & Zhu, W. 2001. Biological conversion of hydrogen sulphide to elemental sulphur in a fixed-film continuous flow photo-reactor. *Water Research*. 35(15):3605–3610.

Henshaw, P., Bewtra, J., & Biswas, N. 1998. Hydrogen Sulfide Conversion to Elemental Sulfur in a Suspended-Growth Continuous Stirred Tank Reactor Using *Chlorobium limicola*. *Water Research.* 32:1769-1778.

Hessler, T., Harrison, S.T.L., & Huddy, R.J., 2018. Stratification of microbial communities throughout a biological sulphate reducing up- flow anaerobic packed bed reactor, revealed through 16S metagenomics. *Research Microbiology.* 169(10): 543-551.

Holkenbrink, C., Barbas, S.O., Mellerup, A., Otaki, H., & Frigaard, N.U. 2011. Sulfur globule oxidation in green sulfur bacteria is dependent on the dissimilatory sulphite reductase system. *Microbiology.* 157(4):1229-39.

Hurse, T., & Keller, J. 2004. Performance of a substratum-irradiated photosynthetic biofilm reactor for the removal of sulfide from wastewater. *Biotechnology and Bioengineering.* 87(1):14–22.

Hussain, A., Hasan, A., Javid, A., & Qazi, J.I. 2016. Exploited application of sulphate-reducing bacteria for concomitant treatment of metallic and non-metalic wastes: a mini review. *3 Biotech.* 6:119.

Hutton, B., Kahan, I., Naidu, T., & Gunther, P. 2009. Operating and maintenance experience at the eMalahleni water reclamation plant. *Proceedings of the International Mine Water Conference.* 19-23rd October 2009. Pretoria, South Africa.145-430.

International Network for Acid Prevention (INAP). 2014. Global Acid Rock Drainage Guide. Available: http://www.gardguide.com. [2019, January, 12].

Ito, T., Okabe, S., Satoh, H., & Watanabe, Y. 2002. Successional Development of Sulphate-Reducing Bacterial Populations and Their Activities in a Wastewater Biofilm Growing under Microaerophilic Conditions. *Applied and Environmental Microbiology.* 68:1392–1402.

Janssen A.J.H., Ruitenberg, R., & Buisman, C.J.N. 2001. Industrial application of new sulphur biotechnology, *Water Science and Technology.* 44(8):85-90.

Janssen, A.J.H., Arena, B.J. & Kijlstra, S., 2000. New developments of the thiopaq process for the removal of H_2S from gaseous streams. *Preprints of Sulphur Conference.* 29:179–187.

Janssen, A.J.H., Lettinga, G., & de Keizer, A. 1999. Removal of hydrogen sulphide from wastewater and waste gases by biological conversion to elemental sulphur. Colloidal and interfacial aspects of biologically produced sulphur particles. *Colloids and Surfaces A: Physicochemical and Engineering Aspects.* 151:389-397.

Jeathon, C., L'Haridon, S., Cueff, V., Banta, A., Reysenbach, A.L., & Prieur, D. 2002. *Thermosulfobacterium hydrogeniphilum* sp. nov., a thermophilic chemolithoautotrophic, sulphate-reducing bacterium isolated from a deep sea hydrothermal vent at Guaymas Basin, and emendation of the genus Thermodesulfobacterium. *International Journal of Systematic and Evolutionary Microbiology.* 52:765-772.

Jensen, A. B., & Webb, C. 1995. Treatment of H_2S-containing gases: a review of microbiological alternatives. *Enzyme and Microbial Technology.* 17:2–10.

Johnson, A.D. 2010. An extended IUPAC nomenclature code for polymorphic nucleic acids. *Bioinfomatics.* 26(10):1386-1389.

Johnson, D.B. 1995. Acidophilic microbial communities: candidates for bioremediation of acidic mine effluents. *International Biodeterioration and Biodegradation.* 35(1-3): 41-58.

Johnson, D.B. 2000. Biological removal of sulfurous compounds from inorganic wastewaters. In: *Environmental technologies to treat sulfur pollution: principles and engineering.* P.N.L. Lens & L.W. Hulshoff Pol, Eds. London: IWA Publishing. 175–205.

Johnson, D.B., & Hallberg, K.B. 2005. Acid mine drainage remediation options: a review. *The Science of the Total Environment.* 338(1-2):3–14.

Jong, T., & Parry, D.L. 2003. Removal of sulphate and heavy metals by sulphate reducing bacteria in short-term bench scale up-flow anaerobic packed bed reactor runs. *Water Research.* 37(14): 3379–3389.

Jovel, J., Patterson, J., Wang, W., Hotte, N., O'Keefe, S., Mitchel, T., Perry, T., Kao, D., Mason, A. L., Madsen, K. L., & Wong, G. K. 2016. Characterization of the Gut Microbiome Using 16S or Shotgun Metagenomics. *Frontiers in microbiology.* 7, 459.

Kabdasli, I., Tünay, O., & Orhon, D. 1995. Sulphate removal from indigo dyeing textile wastewaters. *Water Science and Technology.* 32(12):21-27.

Kaksonen, A. 2004. The performance, kinetics and microbiology of sulfidogenic fluidised- bed reactors treating acidic metal- and sulphate-containing wastewater. Ph.D. Thesis. Tampere University of Technology, Finland.

Kaksonen, H., & Puhakka, J. 2007. Sulphate Reduction Based Bioprocesses for the Treatment of Acid Mine Drainage and the Recovery of Metals. *Engineering in Life Sciences*. 7(6):541–564.

Kantachote, D., Charernjiratrakul, W., Noparatnaraporn, N., & Oda, Kohei. 2008. Selection of sulfur oxidising bacterium for sulfide removal in sulphate rich wastewater to enhance biogas production. *Electronic Journal of Biotechnology*. 11(2):11.

Kelsall, G.H., & Thompson, I. 1993. Redox chemistry of H_2S oxidation by the British Gas Stretford process Part V: Aspects of the process chemistry. *Journal of Applied Electrochemistry*. 23(5):427-434.

Keweloh, H., Heipieper, H.J., & Rehm, H.J. 1989. Protection of bacteria against toxicity of phenol by immobilisation in calcium alginate. *Applied Microbiology and Biotechnology*. 31:383-389.

Khalekuzzaman, M., Hasan, M., Haque, R., & Alamgir, M. 2018. Hydrodynamic performance of a hybrid anaerobic baffled reactor (HABR): effects of number of chambers, hydraulic retention time, and influent temperature. *Water Science Technology*. 78(3-4):968-981.

Kim, B.W., Kim, I.K., & Chang, H.N. 1990. Bioconversion of hydrogen sulphide by free and immobilised cells of *Chlorobium thiosulfatophilum*. *Biotechnology Letters*. 12:381-386.

Kleinjan, W.E., de Keizer, A., & Janssen, A.J.H. 2005. Equilibrium of the reaction between dissolved sodium sulfide and biologically produced sulfur. *Colloids and Surfaces. B, Biointerfaces*. 43(3–4):228–37.

Kobayashi, K. 2007. Bacillus subtilis pellicle formation proceeds through genetically defined morphological changes. *Journal of Bacteriology*. 189(13):4920-4931.

Koizumi, Y., Takii, S., Nishino, M., & Nakajima, T. 2003. Vertical distribution of sulphate reducing bacteria and methane-producing achaea quantified by oligonucleotide probe hybridisation in the profundal sediment of a mesotrophic lake. *FEMS Microbiology Ecology*. 44:101-108.

Kolmert, A., & Johnson, B.D. 2001. Remediation of acidic waste waters using immobilised, acidophilic sulphate-reducing bacteria. *Journal of Chemical Technology & Biotechnology*. 76(8):836–843.

Kovooru, L., Behera, A.K., Alexander, A., & Vuppu, S. 2013. Detection and removal of hydrogen sulphide gas from food sewage water collected from vellore. *Der Pharmacia Lettre.* 5(3):163-169.

Kumar, S., Stecher, G., Li, M., Knyaz, C., & Tamura, K. 2018. MEGA X: Molecular Evolutionary Genetics Analysis across computing platforms. *Molecular Biology and Evolution.* 35:1547-1549.

Kushkevych, I.V. 2016. Dissimilatory sulphate reduction in the intestinal sulphate-reducing bacteria. *Studia Biologica.* 10(1):197-228.

Lens, P., Vallero, M., Esposito, G., & Zandvoort, M. 2003. Perspectives of sulphate reducing bioreactors in environmental biotechnology. *Reviews in Environmental Science & Biotechnology.* 1:311–325.

Lens, P.N.L., Visser, A., Janssen, J.H., Hulshoff Pol, L.W., & Lettinga, G. 2010. Biotechnological Treatment of Sulphate-Rich Wastewaters. *Critical Reviews in Environmental Science and Technology.* 28(1):41-88.

Lens, P.N.L., Visser, A.J.H., L.W. Janssen, Hulshoff Pol L.W., & Lettinga, G. 1998. Biotechnological Treatment of Sulphate-Rich Wastewaters. *Critical Reviews in Environmental Science and Technology.* 28(1):41-88.

Lesnik, K.L., & Liu, H. 2014. Establishing a core microbiome in acetate-fed microbial fuel cells. *Applied Microbiology and Biotechnology.* 98(9):4187-96.

Lewis, A.E., Lahav, O., & Lowenthal, R.E. 2000. Chemical considerations of sulphur recovery from acid mine drainage. Proceedings of the Water Institute of South Africa Biennial Conference.

Li, W., Fu, L., Niu, B., Wu, S., & Wooley, J. 2012. Ultrafast clustering algorithms for metagenomic sequence analysis. *Briefings Bioinformormatics.* 13:656–68.

Liamleam, W., Annachhatre, A.P. 2007. Electron donors for biological sulfate reduction. *Biotechnology Advances.* 25(5):452-463

Logue, J.B., Findlay, S.E.G., & Comte, J. 2015. Editorial: Microbial Responses to Environmental Changes. *Frontiers in Microbiology.* 6:1–4.

Lozupone, C., & Knight, R. 2005. UniFrac: a New Phylogenetic Method for Comparing Microbial Communities. *Applied Environmental Microbiology.* 71:8228–35.

Luther, G.W., Findlay, A.J., Macdonald, D.J., Owings, S.M., Hanson, T.E., Beinart, R.A., & Girguis, P.R. 2011. Thermodynamics and kinetics of sulfide oxidation by oxygen: a look at inorganically controlled reactions and biologically mediated processes in the environment. *Frontiers in Microbiology.* 2(62):1-9.

Lyew, D., & Sheppard, J.D. 1997. Effects of physical parameters of a gravel bed on the activity of sulphate-reducing bacteria in the presence of acid mine drainage, *Journal of Chemical Technology and Biotechnology.* 70:223–230.

Madigan, M., & Jung, D. 2008. An Overview of Purple Bacteria: Systematics, Physiology, and Habitats. In *The Purple Phototrophic Bacteria.* Hunter, C.N., Daldal, F., Thurnauer, M.C. & Beatty, J.T., Eds. Springer. 1–15.

Magoč, T., & Salzberg, S.L. 2011. FLASH: Fast length adjustment of short reads to improve genome assemblies. *Bioinformatics.* 27:2957–63.

Magot, M., Ravot, G., Campaignolle, X., Ollivier, B., Patel, B.K.C., Fardeau, M.L., Thomas, P., Crolet, J.L., & Garcia, J.L. 1997. *Dethiosulphovibrio peptidovorans* gen. nov., sp. nov., a new anaerobic slightly halophilic, thiosulphate reducing bacterium from corroding offshore oil wells. *International Journal of Systematic Bacteriology.* 47(3):818-824.

Maillacheruvu, K.Y., & Parkin, G.F. 1996. Kinetics of growth, substrate utilisation and sulfide toxicity for propionate, acetate, and hydrogen utilisers in anaerobic systems, *Water Environmental Research.* 68:1099–1106.

Maillacheruvu, K.Y., Parkin, G.F., Peng, C.Y., Kuo, W.C., Oonge, Z.I., & Lebduschka, V. 1993. Sulfide Toxicity in Anaerobic Systems Fed Sulphate and Various Organics. *Water Environment Research.* 65(2):100-109.

Mazidji, C.N., Koopman, B., Bitton, G., & Neita, D. 1992. Distinction between heavy metal and organic toxicity using EDTA chelation and microbial assays. *Environmental Toxicology and Water Quality.* 7:339–353.

McCarthy, T.S. 2011. The impact of acid mine drainage in South Africa. *South African Journal of Science.* 107(5/6):1–7.

McDonald, H. 2007. The effect of sulfide inhibition and organic shock loading on anaerobic biofilm reactors treating a low-temperature, high-sulphate wastewater. P.hD thesis. University of Iowa.

Mendez, R., Lema, J.M., & Soto, M. 1995. Treatment of seafood-processing wastewater in mesophilic and thermophilic anaerobic filters. *Water Environment Research*. 67(1):33-45.

Middelburg, J.J. 2000. The geochemical sulphur cycle. In *Environmental technologies to treat sulphur pollution: Principles and Engineering*. P.N.L. Lens & L.W. Hulsoff Pol, Eds. London: IWA Publishing. 33-45.

Molwantwa, J.B. 2008. Floating sulphur biofilms: Structure function and biotechnology. PhD thesis, Rhodes University, Grahamstown.

Molwantwa, J.B., & Rose, P.D. 2013. Development of a Linear Flow Channel Reactor for sulphur removal in acid mine wastewater treatment operations. *Water SA*. 39(5):649–654.

Molwantwa, J.B., Coetser, S.E., Heath, R., Pulles, W., Bowker, M., & Rose, P.D., 2010. *Monitoring, evaluation and long term verification of the passive water treatment plant at VCC*. (WRC report; no. 1623/1/10). South Africa: Water Research Commission.

Mooruth, N. 2013. An investigation towards passive treatment solutions for the oxidation of sulphide and subsequent removal of sulphur from acid mine water. PhD thesis. University of Cape Town, South Africa.

Moosa, S., & Harrison, S.T.L. 2006. Product inhibition by sulphide species on biological sulphate reduction for the treatment of acid mine drainage. *Hydrometallurgy*. 83:214- 222.

Moosa, S., Nemati, M., & Harrison, S.T.L. 2002. A kinetic study on anaerobic reduction of sulphate, Part I: Effect of sulphate concentration. *Chemical Engineering Science*. 57:2773–2780.

Moosa, S., Nemati, M., & Harrison, S.T.L. 2005. A kinetic study on anaerobic reduction of sulphate, part II: incorporation of temperature effects in the kinetic model. *Chemical Engineering Science*. 60(13):3517–3524.

Muyzer, G., & Stams, A.J.M. 2008. The ecology and biotechnology of sulphate-reducing bacteria. *Nature Reviews Microbiology*. 6:441-454.

Muyzer, G., Kuenen, J.G., & Robertson, L.A. 2013. In *The Prokaryotes – Prokaryotic Physiology and Biochemistry*, E. Rosenberg, E. F. DeLong, S. Lory, E. Stackebrandt, F. Thompson, Eds. Berlin, Heidelberg: Springer-Verlag. 555–588.

Nagpal, S., Chuichulcherm, S., & Livingston, A. 2000. Ethanol utilisation by sulphate- reducing bacteria: An experimental and modelling study. *Biotechnology and Bioengineering*. 70:533-543.

Nalcaci, O., Böke, N., & Ovez, B. 2011. Potential of the bacterial strain Acidovorax avenae subsp. avenae LMG 17238 and macro algae Gracilaria verrucosa for denitrification. *Desalination*. 274:44-53.

Neculita, C.M., Zagury, G.J., & Bussière, B. 2008. Effectiveness of sulphate-reducing passive bioreactors for treating highly contaminated acid mine drainage: II. Metal removal mechanisms and potential mobility. *Applied Geochemistry*. 23(12):3545–3560.

Neculita, C.M., Zagury, G.J., & Bussière, B., 2007. Passive treatment of acid mine drainage in bioreactors using sulphate-reducing bacteria: critical review and research needs. *Journal of Environmental Quality*. 36:1-16.

Nielsen, A.H., Hvitved-Jacobsen, T., & Vollertsen, J. 2005. Kinetics and stoichiometry of sulfide oxidation by sewer biofilms. *Water Research*, 39(17):4119–25.

Nielsen, G., Hatam, I., Abuan, K.A., Janin, A., Coudert, L., Blais, J.F., & Baldwin, S.A. 2018. Semi-passive in-situ pilot scale bioreactor successfully removed sulphate and metals from mine impacted water under subarctic climatic conditions. *Water Research*. 140:268–279.

Ntobela, N., & Chibwan, F., 2016. Study of the hydrodynamics of a linear flow channel reactor for application in the bioremediation of acid rock drainage. BEng(Hons) dissertation. University of Cape Town, South Africa.

O'Flaherty, V., & Colleran, E. 1998. Effect of sulphate addition on volatile fatty acid and ethanol degradation in an anaerobic hybrid reactor. I: process disturbance and remediation. *Bioresource Technology*. 68:101-107.

Oyekola, O.O. 2008. An investigation into the relationship between process kinetics and microbial community dynamics in a lactate-fed sulphidogenic CSTR as a function of residence time and sulphate loading. PhD Thesis. University of Cape Town, South Africa.

Oyekola, O.O., Harrison, S.T.L., & van Hille, R.P. 2012. Effect of culture conditions on the competitive interaction between lactate oxidisers and fermenters in a biological sulphate reduction system. *Bioresoure Technology*. 104:616-621.

Oyekola, O.O., van Hille, R.P., & Harrison, S.T.L. 2010. Kinetic analysis of biological sulphate reduction using lactate as carbon source and electron donor: Effect of sulphate concentration. *Chemical Engineering Science*. 65(16):4771–4781.

Oyekola, O.O., van Hille, R.P., & Harrison, S.T.L., 2009. Study of anaerobic lactate metabolism under biosulfidogenic conditions. *Water Research*, 43(14):3345–54.

Paytubi, S., Cansado, C., Madrid, C., & Balsalobre, C. 2017. Nutrient Composition Promotes Switching between Pellicle and Bottom Biofilm in Salmonella. *Frontiers in Microbiology.* 8:1–12.

Pokorna, D., & Zabranska, J. 2015. Sulfur-oxidising bacteria in environmental technology. *Biotechnology Advances.* 33:1246.

Poretsky, R., Rodriguez, L.M., Luo, C., Tsementzi, D., & Konstantinidis, K.T. 2014. Strengths and Limitations of 16S rRNA Gene Amplicon Sequencing in Revealing Temporal Microbial Community Dynamics. *PLoS ONE.* 9.

Postgate, J.R. 1984. The Sulphate Reducing Bacteria, 2nd ed. Cambridge, UK: University Press. 9-23.

Pruden, A., Messner, N., Pereyra, L., Hanson, R.E., Hiibel, S.R., & Reardon, K.F. 2007. The effect of inoculum on the performance of sulphate-reducing columns treating heavy metal contaminated water. *Water Research.* 41:904–914.

Pulles, W., & Heath, R. 2009. The evolution of passive mine water treatment technology for sulphate removal. *Proceedings of the Water Institute of Southern Africa & International Mine Water Association conference.* 19-23 October 2009. Pretoria, South Africa. 2-15.

Ramel, F., Brasseur, G., Pieulle, L., Valette, O., Hirschler-Réa, A., Fardeau, M.L., & Dolla, A. 2015. Growth of the obligate anaerobe *desulfovibrio vulgaris* hildenborough under continuous low oxygen concentration sparging: impact of the membrane-bound oxygen reductases. *PLoS ONE.* 10(4):1-17.

Rao, A.G., Ravichandra, P., Joseph, J., Jetty, A., & Sarma, P.N. 2007. Microbial conversion of sulfur dioxide in flue gas to sulfide using bulk drug industry wastewater as an organic source by mixed cultures of sulphate reducing bacteria. *Journal of Hazardous Materials.* 147(3):718-725.

Raskin, L., Rittmann, B.E., & Stahl, D.A. 1996. Competition and coexistence of sulphate reducing and methanogenic populations in anaerobic biofilms. *Applied and Environmental Microbiology.* 62:3847-3857.

Rein, N. B. 2002. Biological Sulphide Oxidation in Heterotrophic Environments. MSc thesis. Rhodes University, South Africa.

Reis, M.A.M., Almeida, J.S., Lemos, P.C. & Carrondo, M.J.T. 1992. Effect of hydrogen sulphide on growth of sulphate-reducing bacteria. *Biotechnology and Bioengineering.* 40:593-600.

Robador, A., Muller A.L., Sawicka J.E., Berry, D., Hubert, C.R., Jorgensen, B.B., & Bruchert, V. 2016. Activity and community structures of sulphate- reducing microorganisms in polar, temperate and tropical marine sediments. *ISME Journal.* 10(4):796–809.

Roman, H. 2004. The degradation of lignocellulose in a biologically-generated sulphidic environment. PhD thesis, Rhodes University.

Rose, P. 2013. Long-term sustainability in the management of acid mine drainage wastewaters – development of the Rhodes BioSURE Process. *Water SA.* 39(5):583–592.

Rose, P.D. 2002. Salinity, sanitation and sustainability: A study in environmental biotechnology and integrated wastewater beneficiation in South Africa. (WRC report; no. TT 187/02). South Africa: Water Research Commission. ISBN: 1-86845-884.

Rubio-Rincon, F., Lopez-Vazquez, C., Welles, L., van den Brand, T., Abbas, B., van Loosdrecht, M., & Brdjanovic, D. 2017. Effects of electron acceptors on sulphate reduction activity in activated sludge processes. *Applied Microbiology and Biotechnology.* 101(15):6229-6240.

Rückert, C. 2016. Sulphate reduction in microorganisms — recent advances and biotechnological applications. *Current Opinion in Microbiology.* 33:140–146.

Rudi, K., Zimonja, M., Trosvik, P., & Naes, T. 2007. Use of multivariate statistics for 16S rRNA gene analysis of microbial communities. *International Journal of Food Microbiology.* 120:95–99.

Saez-Navarrete, C., Zamorano, A., Ferrada, C., & Rodriguez-Cordova, L. 2009. Sulphate reduction and biomass growth rates for *Desulfobacterium autotrophicum* in yeast extract-supplemented media at 38°C. *Desalination.* 348(1):

Sanchez, R.F., Cordoba, P., & Sineriz, F., 1997. Use of the USAB for the anaerobic treatment of stillage from sugar-cane molasses. *Biotechnology and Bioengineering.* 27(12):1710–1716.

Sánchez-Andrea, I., Sanz, J.L., Bijmans, M.F.M., & Stams, A.J.M., 2014. Sulphate reduction at low pH to remediate acid mine drainage. *Journal of Hazardous Material.* 269(3):98–109.

Santos, A.L., & Johnson, D.B. 2018. Design and application of a low pH upflow biofilm sulphidogenic bioreactor for recovering transition metals from synthetic waste water at a Brazilian copper mine. *Frontiers in Microbiology*. 9(2051):1-11.

Sarti, A., & Zaiat, M. 2011. Anaerobic treatment of sulphate-rich wastewater in an anaerobic sequential batch reactor (AnSBR) using butanol as the carbon source. *Journal of Environmental Management*. 92(6):1537-1541.

Sass, A. M., Eschemann, A., Kuhl, M., Thar, R., Sass, H., & Cypionka, H. 2002. Growth and chemosensory behavior of sulphate-reducing bacteria in oxygen-sulfide gradients. *FEMS Microbiology Ecology*. 40(1):47–54.

Sato, Y., Hamai, T., Hori, T., Habe, H., Kobayashi, M., & Sakata, T., 2017. Year-round performance of a passive sulphate-reducing bioreactor that uses rice bran as an organic carbon source to treat acid mine drainage. *Mine Water and the Environment*. 37(3):586-594.

Sawicka, J. E., Jorgensen, B.B., & Bruchert, V. 2012. Temperature characteristics of bacterial sulphate reduction in continental shelf and slope sediments. *Biogeoscience*. 9:3425–3435.

Shen, Y., & Buick, R. 2004. The antiquity of microbial sulphate reduction. *Earth-Science Reviews*. 64(3-4):243-272.

Sheoran, A., Sheoran, V., & Choudhary, R.P. 2010. Bioremediation of acid-rock drainage by sulphate-reducing prokaryotes: A review. *Minerals Engineering*. 23(14):1073–1100.

Silva, A.J., Hirasawa, J.S., Varesche, M.B., Foresti, E., & Zaiat, M. 2006. Evaluation of support materials for the immobilisation of sulphate-reducing bacteria and methanogenic archaea. *Anaerobe*. 12(2):93–8.

Skousen, J., Zipper, C.E., Rose, A., Ziemkiewicz, P.F., Nairn, R., McDonald, L.M., & Kleinmann, R.L. 2017. Review of passive systems for acid mine drainage treatment. *Mine Water and the Environment*. 36(1):133-153.

Song, Y.C., Piak, B.C., Shin, H.S., & La, S.J. 1998. Influence of electron donor and toxic materials on the activity of sulphate reducing bacteria for the treatment of electroplating wastewater. *Water Science and Technology*. 38(4–5):187-194.

Sposob, M., Bakke, R., & Dinamarca, C. 2017. Modeling N/S ratio and temperature influences on simultaneous biological denitrification and sulfide oxidation. *Proceedings of the 58th SIMS conference*. 25 – 27 September 2017. Reykjavik, Iceland.

Stams, A. J. M., Huisman, J., Garcia Encina, P. A., & Muyzer, G. 2009. Citric acid wastewater as electron donor for biological sulphate reduction. *Applied Microbiology and Biotechnology*. 83:957–963.

Sun, J., Dai, X., Liu, Y., Peng, L., & Ni, B.J. 2017. Sulfide removal and sulfur production in a membrane aerated biofilm reactor: model evaluation. *Chemical Engineering Journal*. 309:454–462.

Syed, M., Soreanu, G., Falletta, P., & Béland, M. 2006. Removal of hydrogen sulfide from gas streams using biological processes - A review. *Canadian Biosystems Engineering*. 48(2):1-14.

Tian, H., Gao, P., Chen, Z., Li, Y., Li, Y., Wang, Y., Zhou, J., Li, G., & Ma, T. 2017. Compositions and abundances of sulphate-reducing and sulfur-oxidising microorganisms in water-flooded petroleum reservoirs with different temperatures in China. *Frontiers in Microbiology*. 8:1–14.

Tichy, R., Lens, P., Grotenhuis, J.T.C., & Bos, P. 2010. Solid-state reduced sulfur compounds: environmental aspects and bioremediation. *Critical Reviews in Environmental Science and Technology*. 28:1-40.

Tsukamoto, T., & Miller, G. 1999. Methanol as a carbon source for microbiological treatment of acid mine drainage. *Water Research*. 33(6):1365–1370.

Tsukamoto, T.K., Killion, H.A., & Miller, G.C. 2004. Column experiments for microbial treatment of acid mine drainage: low-temperature, low-pH and matrix investigations. *Water Research*. 38:1405–1418.

Utgikar, V.P., Harmon, S.M., Chaudhary, N., Tabak, H.H., Govind, R., & Haines, J.R. 2002. Inhibition of sulphate-reducing bacteria by metal sulfide formation in bioremediation of acid mine drainage. *Environmental Toxicology*. 17:40–48.

Utgikar, V.P., Tabak, H.H., Haines, J.R., & Govind, R. 2003. Quantification of toxic and inhibitory impact of copper and zinc on mixed cultures of sulphate-reducing bacteria. *Biotechnology and Bioengineering*. 82(3):306–12.

van den Bosch, P. L. F., Sorokin, D. Y., Buisman, C. J. N., & Janssen, A. J. H. 2008. The effect of pH on thiosulfate formation in a biotechnological process for the removal of hydrogen sulfide from gas streams. *Environmental Science and Technology.* 42(7), 2637-2642.

van den Brand, T.P.H., Roest, K., Chen, G.H., Brdjanovic, D., & van Loosdrecht, M.C.M. 2016. Adaptation of Sulphate-Reducing Bacteria to Sulfide Exposure. *Environmental Engineering Science.* 33:242–249.

van Hille, R.P., & Mooruth, N., 2014. *Investigation of carbon flux and sulphide oxidation kinetics during passive biotreatment of mine mater.* (WRC report; no. 2139/1/13). South Africa: Water Research Commission. ISBN: 978-1-4312-0491-5.

van Hille, R.P., Boshoff, G.A., Rose, P.D., & Duncan, J.R., 1999. A continuous process for the biological treatment of heavy metal contaminated acid mine water. *Resources, Conservation and Recycling.* 27(1):157-167.

van Hille, R.P., Marais T.S., & Harrison, S.T.L. 2015. Biomass Retention and Recycling to Enhance Sulphate Reduction Kinetics. *Proceedings of the 10th ICARD & IMWA Annual Conference on Agreeing on Solutions for More Sustainable Mine Water Management.* 21-24 April 2015. Santiago, Chile. 250-275.

van Hille, R.P., Motleleng, L., Van Wyk, N., Huddy, R., Smart, M., & Harrison, S.T.L. 2015. *Development of a toolkit to enable quantitative microbial ecology studies of sulphate reducing and sulphide oxidising systems.* (WRC report; no. 2109/1/14). South Africa: Water Research Commission. ISBN: 978-1-4312-0604-9.

van Hille, R.P., van Wyk, N., Motleleng, L., & Mooruth, N., 2011. Lessons in passive treatment: Towards efficient operation of a sulphate reduction-sulphide oxidation system. *Proceedings of the 11th IMWA Congress on Mine Water: Managing the Challenges.* 4-11 September 2011. Aachen, Germany. 491 – 495.

Vasquez, Y., Escobar, M.C., Neculita, C.M., Arbeli, Z., & Roldan, F. 2016. Biochemical passive reactors for treatment of acid mine drainage: Effect of hydraulic retention time on changes in efficiency, composition of reactive mixture, and microbial activity. *Chemosphere.* 153:244–253.

Vasquez, Y., Escobar, M.C., Saenz, J.S., Quiceno-Vallejo, M.F., Neculita, C.M, Arbelib, Z., & Roldanb, F. 2018. Effect of hydraulic retention time on microbial community in biochemical passive reactors during treatment of acid mine drainage. *Bioresource Technololgy.* 247:624–632.

Vikromvarasiri, N., Champreda, V., Boonyawanich, S., & Pisutpaisal, N. 2017a. Hydrogen sulfide removal from biogas by biotrickling filter inoculated with *Halothiobacillus neapolitanus*. *International Journal of Hydrogen Energy.* 1-9.

Vikromvarasiri, N., Juntranapaporn, J., & Pisutpaisal, N. 2017b. Performance of *Paracoccus pantotrophus* for H_2S removal in biotrickling filter. *International Journal of Hydrogen Energy.* 42(45): 27820-27825.

Visser, A. 1995. The anaerobic treatment of sulphate containing wastewater. Ph.D thesis. Wageningen Agricultural University, Wageningen, Netherlands.

White, C., & Gadd, G.M. 1996. Mixed sulphate-reducing bacterial cultures for bioprecipitation of toxic metals: factorial and response-surface analysis of the effects of dilution rate, sulphate and substrate concentration. *Microbiology.* 142:2197-2205.

WHO. 2004. Sulphate in drinking water. Background document for the development of WHO guidelines for drinking-water quality WHO/SDE/WSH/03.04/114, World Health Organisation.

Woese, C.R., Magrum, L.J., Gupta, R., Siegel, R.B., Stahl, D.A., Kop, J., Crawford, N., Brosius, J., Gutell, R., Hogan, J.J., & Noller, F.F. 1980. Secondary structure model for bacterial 16S ribosomal RNA: phylogenetic, enzymatic and chemical evidence. *Nucleic Acids Research.* 8(10):2275-2293.

Xu, X.J., Chen, C., Wang, A. J., Fang, N., Yuan, Y., Ren, N.Q., & Lee, D.J. 2012. Enhanced elementary sulfur recovery in integrated sulphate-reducing, sulfur-producing rector under micro-aerobic condition. *Bioresource Technology.* 116:517–521.

Yang, B., Wang, Y., & Oian, P. 2016. Sensitivity and correlation of hypervariable regions in 16S rRNA genes in phylogenetic analysis. *BMC Bioinformatics.* 17(135):1-8.

Zagury, G.J., Kulnieks, V.I., & Neculita, C.M. 2007. Characterisation and reactivity assessment of organic substrates for sulphate-reducing bacteria in acid mine drainage treatment. *Chemosphere.* 64:944–954.

Zarei, O., Dastmalchi, S., & Hamzeh-Mivehroud M. 2016. A Simple and Rapid Protocol for Producing Yeast Extract from *Saccharomyces cerevisiae* Suitable for Preparing Bacterial Culture Media. *Iran Journal of Pharmacy Research.* 15(4):907–913.

Zhang, M., & Wang, H. 2014. Organic wastes as carbon sources to promote sulphate reducing bacterial activity for biological remediation of acid mine drainage. *Minerals Engineering*. 69:81–90.

Zhang, Y., Wang, X., Zhen, Y., Mi, T., He, H., & Yu, Z. 2017. Microbial diversity and community structure of sulphate-reducing and sulfur-oxidising bacteria in sediment cores from the East China Sea. *Frontiers in Microbiology*. 8:1–17.